Technologies for the Electrification of Land and Sea Transportation

Other related titles:

You may also like

- PBRN007 | Miller | Propulsion Systems for Hybrid Vehicles, 2nd Edition | 2010
- PBTR016 | Kishor | ICT for Electric Vehicle Integration with the Smart Grid | 2019
- PBPO079 | Lu | Vehicle-to-Grid: Linking electric vehicles to the smart grid | 2015
- PBPO111 | Zobaa | Energy Storage at Different Voltage Levels: Technology, integration, and market aspects | 2018
- PBPO092 | Zobaa | Power Quality in Future Electrical Power Systems | 2017

We also publish a wide range of books on the following topics:
Computing and Networks
Control, Robotics and Sensors
Electrical Regulations
Electromagnetics and Radar
Energy Engineering
Healthcare Technologies
History and Management of Technology
IET Codes and Guidance
Materials, Circuits and Devices
Model Forms
Nanomaterials and Nanotechnologies
Optics, Photonics and Lasers
Production, Design and Manufacturing
Security
Telecommunications
Transportation

All books are available in print via
https://shop.theiet.org or as eBooks via our Digital Library https://digital-library.theiet.org.

IET TRANSPORTATION SERIES 042

Technologies for the Electrification of Land and Sea Transportation

Edited by
Muhyaddin J.H. Rawa, Foad H. Gandoman,
Ziad M. Ali and Shady H.E. Abdel Aleem

The Institution of Engineering and Technology

About the IET

This book is published by the Institution of Engineering and Technology (The IET).

We inspire, inform and influence the global engineering community to engineer a better world. As a diverse home across engineering and technology, we share knowledge that helps make better sense of the world, to accelerate innovation and solve the global challenges that matter.

The IET is a not-for-profit organisation. The surplus we make from our books is used to support activities and products for the engineering community and promote the positive role of science, engineering and technology in the world. This includes education resources and outreach, scholarships and awards, events and courses, publications, professional development and mentoring, and advocacy to governments.

To discover more about the IET please visit https://www.theiet.org/.

About IET books

The IET publishes books across many engineering and technology disciplines. Our authors and editors offer fresh perspectives from universities and industry. Within our subject areas, we have several book series steered by editorial boards made up of leading subject experts.

We peer review each book at the proposal stage to ensure the quality and relevance of our publications.

Get involved

If you are interested in becoming an author, editor, series advisor, or peer reviewer please visit https://www.theiet.org/publishing/publishing-with-iet-books/ or contact author_support@theiet.org.

Discovering our electronic content

All of our books are available online via the IET's Digital Library. Our Digital Library is the home of technical documents, eBooks, conference publications, real-life case studies and journal articles. To find out more, please visit https://digital-library.theiet.org.

In collaboration with the United Nations and the International Publishers Association, the IET is a Signatory member of the SDG Publishers Compact. The Compact aims to accelerate progress to achieve the Sustainable Development Goals (SDGs) by 2030. Signatories aspire to develop sustainable practices and act as champions of the SDGs during the Decade of Action (2020–30), publishing books and journals that will help inform, develop, and inspire action in that direction.

In line with our sustainable goals, our UK printing partner has FSC accreditation, which is reducing our environmental impact to the planet. We use a print-on-demand model to further reduce our carbon footprint.

Published by The Institution of Engineering and Technology, London, United Kingdom

The Institution of Engineering and Technology (the "**Publisher**") is registered as a Charity in England & Wales (no. 211014) and Scotland (no. SC038698).

First published 2025

The Institution of Engineering and Technology
Futures Place
Kings Way, Stevenage
Herts, SG1 2UA
United Kingdom

www.theiet.org

British Library Cataloguing in Publication Data

A catalogue record for this product is available from the British Library

ISBN 978-1-83953-721-9 (hardback)
ISBN 978-1-83953-722-6 (PDF)

Typeset in India by MPS Limited

Cover image: Andriy Onufriyenko/Moment via Getty Images

Contents

About the editors

Muhyaddin J.H. Rawa is a professor with the Department of Electrical and Computer Engineering at King Abdulaziz University, Saudi Arabia, where he is the head of the Smart Grids Research Group. He received a PhD in electrical and electronic engineering from the University of Nottingham in 2014. He has over 7 years of experience at Saudi Electricity Company. He has more than 150 publications and 17 United States Patent and Trademark Office (USPTO) patents.

Foad H. Gandoman is a CTO of the Reliability & Safety Technical Centre, Belgium. He earned his PhD with the Battery Team (BIC) at the Vrije Universiteit Brussel (VUB) in Belgium. He has been project engineer, research assistant, and technical project manager on EU projects such as ASSURED, GHOST, Vision-xEV, and ACHILES. He is the editor of the *International Journal of Electrical Engineering & Education (IJEEE).*

Ziad M. Ali is an associate professor at Aswan University, Egypt, and Prince Sattam bin Abdulaziz University, Saudi Arabia. He received his PhD in 2010 from Kazan State Technical University, Russia. He worked as an assistant professor in the faculty of engineering at Aswan University, Egypt from 2011 to 2016. He was a visitor researcher at Kazan State Energy University (2012–13), and the University of Padova, Italy (2013–14).

Shady H.E. Abdel Aleem is an associate professor at Science Valley Academy, Egypt, and an adjunct associate professor at the Arab Academy for Science, Technology and Maritime Transport. He is also a consultant for ETA Electric Company, Egypt. He was awarded the State Encouragement Award in engineering sciences in 2017, and the medal of distinction from the first class of the Egyptian State Award in 2020.

Chapter 1

Impact of the public fleet electrification on modern electric distribution networks

Omar Abdelrahim[1,2], *Noura A. Nouraldin*[3], *Sherif M. Dabour*[4,5], *Wessam Rahouma*[6] and *Ziad M. Ali*[1,7]

In recent years, electrical fleets within distribution systems have gained significant market share due to their role in conserving natural resources and reducing pollution. As the development of charging infrastructure for electric vehicles (EVs) continues to grow, the impact of electric vehicle stations (EVS) on power quality (PQ) becomes a crucial consideration. EVs, owing to their bidirectional grid-to-vehicle and vehicle-to-grid charging capabilities, pose challenges to the traditional electrical grid. The proliferation of EVs is associated with a corresponding increase in PQ issues within the electrical distribution infrastructure, including concerns such as harmonic distortion, transformer failures, and voltage imbalances.

Furthermore, many municipalities are making the transition from diesel to electric buses for significant cost savings and a reduced carbon footprint. Electric buses, however, differ from other EVs in terms of their operational duties and charging patterns. As a result, planning charging infrastructure requires a comprehensive approach that considers both the transportation network and the power grid. Integrating large-scale electricity-consuming stations, like those used for charging electric buses, imposes substantial and somewhat unpredictable demands on distribution networks.

In addition, the maritime industry is experiencing a shift from mechanical propulsion to electric propulsion, particularly through the utilization of variable frequency drives for motor control. This transition has introduced a

[1]Electrical Engineering Department, Faculty of Engineering, Aswan University, Egypt
[2]Electrical Power Engineering, Egypt-Japan University of Science and Technology (E-JUST), Egypt
[3]Electrical Department, Faculty of Technology and Education, Suez University, Egypt
[4]Department of Electrical Power Engineering, Glasgow Caledonian University, UK
[5]Department of Electrical Power Engineering, Faculty of Engineering, Tanta University, Egypt
[6]Electrical Power and Renewable Energy, University of Doha for Science and Technology, Qatar
[7]College of Engineering at Wadi Addawasir, Prince Sattam Bin Abdulaziz University, Kingdom of Saudi Arabia

notable increase in non-linear loads due to the adoption of electric propulsion in large vessels used for commercial or military purposes.

This chapter aims to examine the impacts and challenges associated with the integration of electric fleet loads into distribution systems. Specifically, it investigates the effects of electric fleet loads on line and transformer overloads, as well as their influence on power quality and voltage stability.

1.1 Introduction

In 2014, the United States found itself grappling with a substantial environmental challenge, as shown in Figure 1.1. Fossil fuel combustion, a dominant force in the nation's energy landscape, was responsible for a staggering 76% of the total carbon dioxide equivalent (CO2-e) emissions, amounting to a formidable 6,870 million metric tons (Mt). To address this issue, electrifying end-use technologies while simultaneously decarbonizing electricity generation has emerged as a viable solution. This solution revolves around the replacement of traditional fossil fuels such as coal, natural gas, and petroleum-based fuels with alternative sources characterized by lower carbon intensity. In doing so, it plays a pivotal role in mitigating emissions originating from fossil fuel combustion across diverse sectors (Table 1.1) [1–4].

It's noteworthy that the carbon intensity of electricity has shown a consistent downward trajectory, a trend driven by factors beyond explicit efforts to lower the grid's carbon footprint. This underscores the considerable potential for emissions

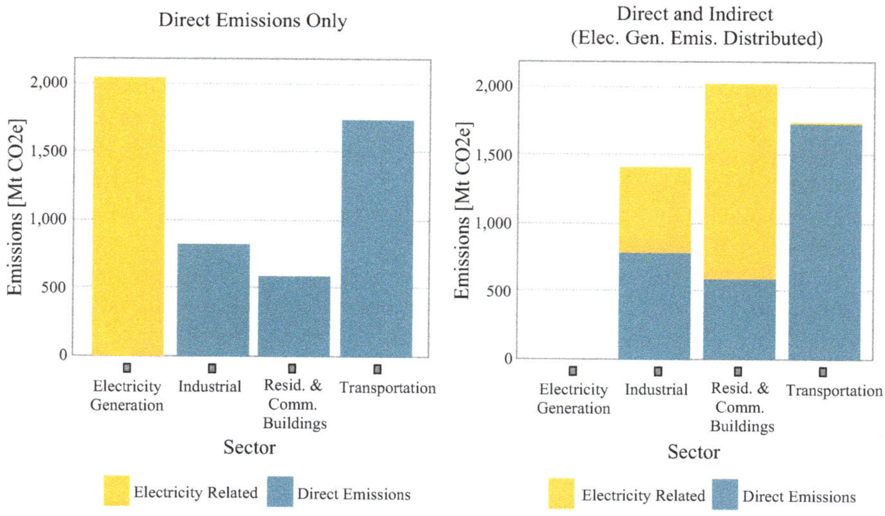

Figure 1.1 2014 GHG emissions from fossil fuel combustion by sector (left) and by sector with emissions associated with electricity generation distributed to end-use sectors (right)

Table 1.1 Challenges of electric vehicle in PQ

Challenges	Description
Power losses	Power loss increases when there are several uncontrolled, single-phase EV charging stations. Distribution transformer losses and overloading are brought on by high EV adoption.
Voltage unbalance	Due to the system's inefficient three-phase distribution of all single-phase demand, single-phase electric vehicle charging big loads creates voltage imbalance.
Voltage fluctuations	As charge rate and penetration rise, the impact increases.
Harmonics	The impact grows when charging rate and penetration grow along with EV charging at random.

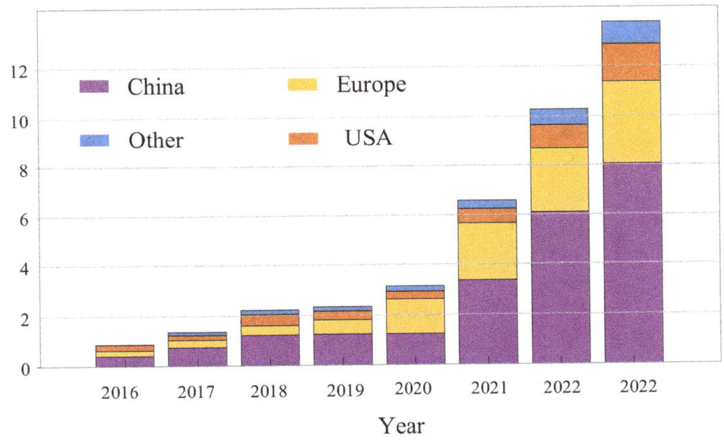

Figure 1.2 Electrical vehicles from 2016 to 2023

reduction through the electrification of specific end uses. In addition, the diminished reliance on fossil fuels brought about by electrification has the potential to curtail domestic coal, oil, and natural gas production, yielding a substantial reduction in upstream emissions. While the analysis given in this chapter primarily focuses on combustion-related emissions, it's essential to acknowledge that these upstream emissions, accounting for 7% of US GHG emissions in 2014, can be substantial (Figure 1.1) [5–11].

Combustion emissions (encompassing CO_2, CH_4, and N_2O) are a major contributor to climate change, and they are distributed across all sectors of the US economy. However, some sectors contribute more to combustion emissions than others. In 2014, the power sector was responsible for 39% of the total US combustion emissions, followed by transportation (34%), industry (15%), and residential and commercial buildings (11%), as shown in Figure 1.2.

Given the substantial share of emissions attributed to the power sector, the decarbonization of electricity generation alone could significantly impact the overall economy-wide emissions. For instance, by reducing the emissions intensity of electricity generation by 80%, a level demonstrated as attainable in various studies, it could result in a reduction of nearly 1,650 Mt of CO_2 (Figure 1.2). This represents 24% of the total US GHG emissions and nearly one-third of combustion emissions (Figure 1.2). However, the combination of grid decarbonization and end-use electrification holds even more potential to address combustion-related GHG emissions (Figure 1.2).

Electrification can also directly address direct combustion emissions tied to end-use services within sectors such as transportation, industry, and buildings. These sectors collectively accounted for 3,158 Mt CO2-e emissions in 2014, constituting 46% of the total US GHG emissions (Figure 1.3).

Among these sectors, transportation contributed to the largest share of direct combustion emissions (1,754 Mt CO2-e), indicating substantial potential for direct emissions reductions through electrification (Figure 1.3). The data further illustrates that emissions in the transportation sector largely result from direct fossil fuel combustion, making electrification, particularly transitioning to electric vehicles (EVs), a promising avenue for achieving significant GHG reductions (Figure 1.3).

The significance of electrification extends to emissions mitigation and the decarbonization of energy supply chains, positioning it as a crucial strategy to achieve net-zero goals. As the adoption of electric technologies increases, the share of electricity in the total final energy consumption is projected to grow from 20% in 2022 to over 27% in 2030 within the Net Zero Emissions by 2050 (NZE) Scenario (Figure 1.4).

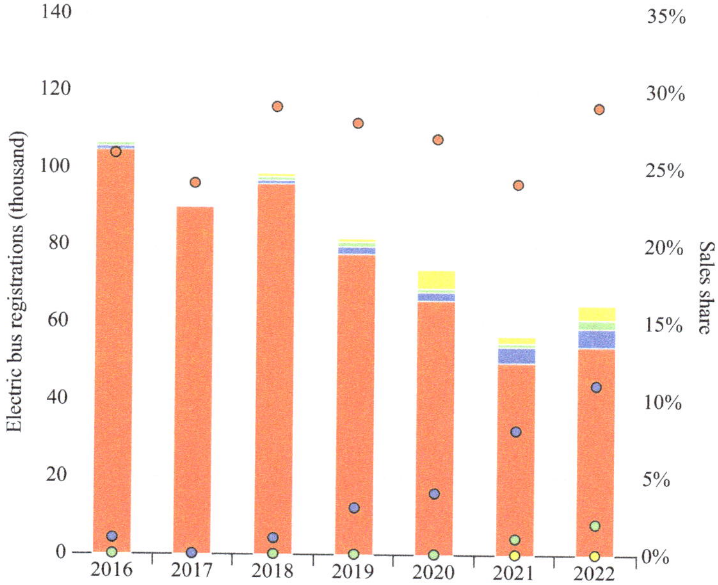

Figure 1.3 Electric bus registrations and sales share by region, 2016–2022

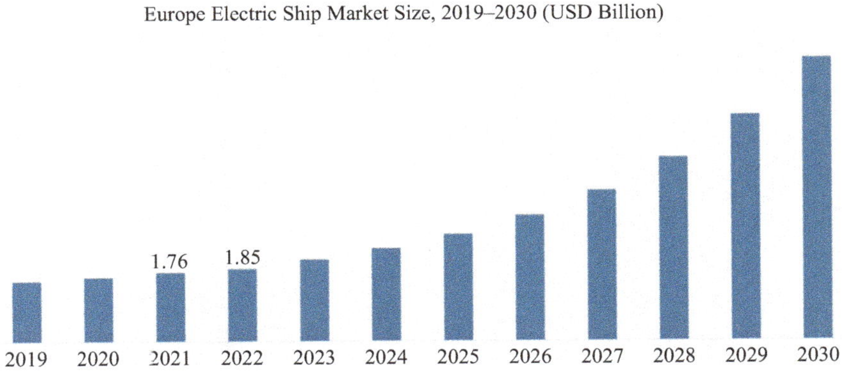

Figure 1.4 Electric ship market size, 2019–2030

While progress has been consistent, accelerating this trend is essential to align with the NZE Scenario's milestones. Strategies like transitioning to electric transport and adopting heat pumps can contribute to achieving this growth (Figure 1.4). The potential for electrification is particularly promising in low-temperature industrial processes such as food drying and beverage production. However, the competitive nature of the market and the longevity of industrial equipment may result in a slower pace of electrification compared to other areas (Figure 1.4) [12–20].

Notable achievements related to electrification worldwide include:

- China exceeded its target with EV sales reaching 29% in 2022, surpassing the planned 20% new energy vehicle sales for 2025.
- Heat pumps outperformed fossil fuel-based heating systems in 2022 in both France and the United States.
- More than half of India's three-wheeler registrations in 2022 were for EVs.
- New Zealand took steps toward cleaner alternatives by prohibiting the installation of new low- and medium-temperature coal boilers in 2021 (Figure 1.5).

Within the Net Zero by 2050 Scenario, the most substantial reductions in emissions due to electrification are projected to occur in the transportation sector (Figure 1.6). The efficiency advantage of electric technologies over fossil fuel-based alternatives, coupled with increased deployment of renewable energy sources, contributes to both emission reduction and the integration of variable renewables into the energy system (Figure 1.6).

Specifically, the light-duty vehicle segment in the road transport sector is anticipated to experience the most significant reductions, with an efficiency-enhanced powertrain alone projected to avoid approximately 1 Gt of CO_2 emissions in 2030 compared to current levels under the NZE Scenario (Figure 1.6). The electrification of space heating also makes a notable contribution. While natural gas boilers currently dominate residential heating, new energy efficiency standards for buildings are expected to elevate the adoption (Figure 1.7).

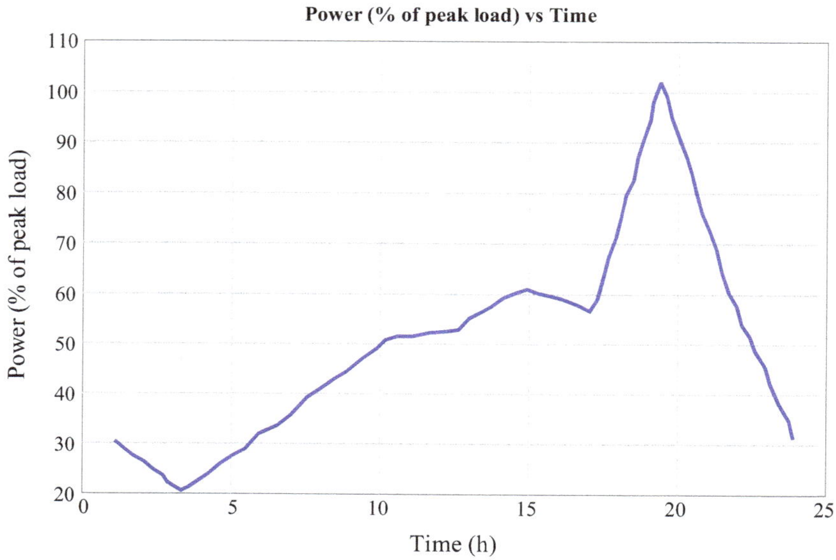

Figure 1.5 Impact of EVs charging on voltage unbalance

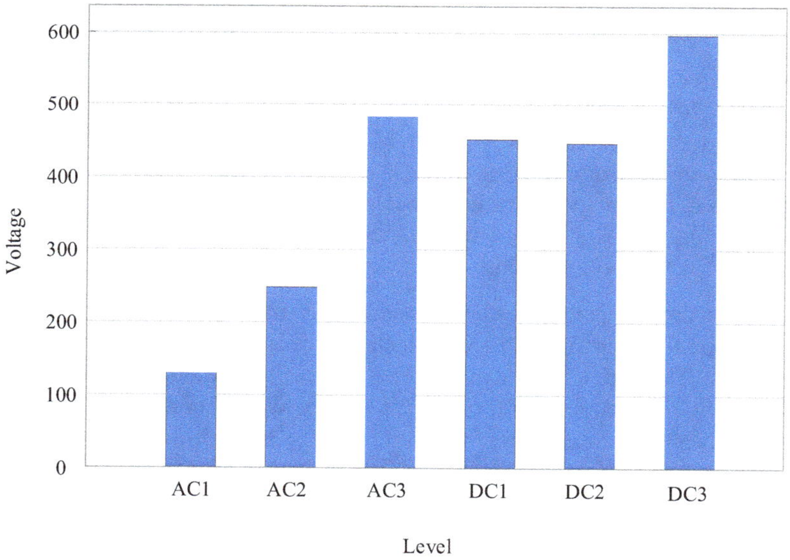

Figure 1.6 AC and DC level of voltage in EVs

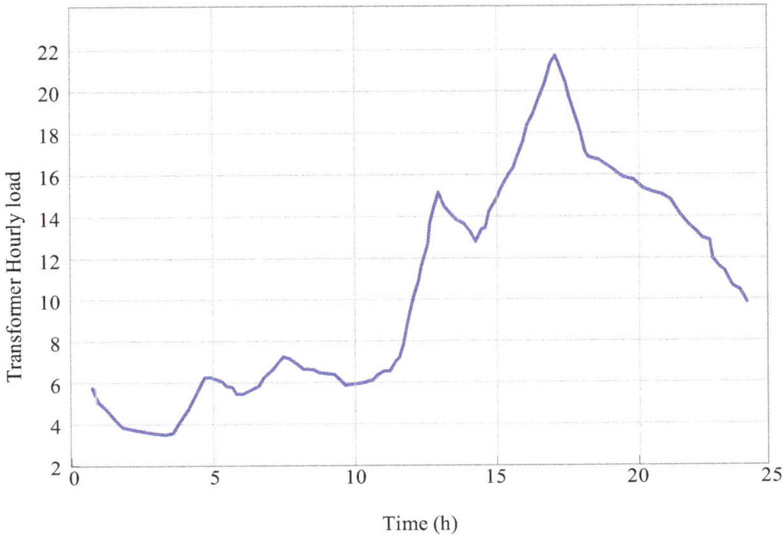

Figure 1.7 Transformer hour load kVA vs time

1.2 Importance of fleet electrification

The transportation sector in the United States and the United Kingdom has now surpassed electricity generation as the largest contributor to greenhouse gas emissions. Meeting decarbonization goals necessitates addressing emissions from this sector.

Shifting from traditional fuels to sustainable alternatives, such as electricity, offers a practical solution. While electric passenger vehicles receive substantial media attention, the potential for fleet electrification, especially in medium- and heavy-duty vehicles (MHDVs), is significant. These MHDVs accounted for nearly a quarter (25%) of US transportation emissions in 2019.

In the United Kingdom, electric refuse trucks are also playing a crucial role as local councils and waste companies strive for net-zero emissions. These battery-powered vehicles offer emission reductions, access to low-emission zones without charges, and exemption from vehicle excise duty. For instance, Manchester City Council and Biffa replaced half of their refuse collection fleet with 27 electric trucks, resulting in approximately 900 tons of carbon emissions saved annually. The City of London Corporation procured five electric trucks to help meet nitrogen dioxide guidelines. Dennis Eagle's eCollect truck, deployed by various councils, exemplifies EV advancements. Rural areas also benefit, with Powys County Council and Dumfries and Galloway Council adopting electric refuse trucks. Furthermore, companies are refurbishing old diesel trucks into electric ones, as seen with Refuse Vehicle Solutions converting a 2015 vehicle for Wrexham County Council. These efforts align with the UK's ambition to achieve net-zero emissions by 2050 [21–24].

1.3 Crucial role in addressing transportation emissions

Addressing emissions from internal combustion engine (ICE) vehicles, which currently dominate the transportation sector, is crucial for mitigating climate change, environmental issues, and transportation equity challenges. The consensus among experts is to electrify transportation and power vehicles with clean electricity. While the feasibility of alternative fuels like hydrogen for specific MHDVs is under debate, several factors, including public policy, fleet adaptability, and economic incentives, point toward an era of extensive fleet electrification [25–27].

1.4 Policy momentum for wider electric fleet adoption

Fifteen states and the District of Columbia, along with support from 37 businesses, have committed to achieving 100% zero-emission MHDV sales by 2050, with intermediate goals of 30% by 2030. Federal support, exemplified by President Biden's pledge to transition the federal fleet to EVs, further fuels the momentum for fleet electrification. This shift is well founded due to the significant contribution of cargo trucks and buses to fuel consumption and emissions, as illustrated in Figure 1.8. It not only reduces pollution but also yields immediate health benefits and aligns with ensuring equitable access to clean public transportation.

1.5 Corporate engagement and rapid fleet electrification

Corporate involvement is accelerating the pace of fleet electrification. A noteworthy example is National Grid's participation in the EV100 initiative, aiming to electrify 5 million vehicles by 2030. Importantly, the transition to electric fleets is projected to gather momentum even without additional policy interventions. According to Bloomberg New Energy Finance (BNEF 2021), commercial MHDV

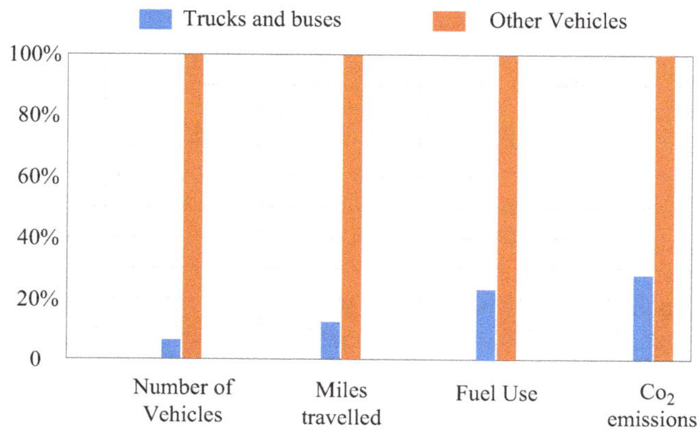

Figure 1.8 Cargo trucks and buses as a proportion of all vehicles and emissions

sales of EVs are anticipated to surpass 10% by 2030 and nearly 30% by 2040. However, achieving net-zero emissions by 2050 may necessitate policies to drive even faster adoption [28–31].

1.6 Varied rates of electrification across vehicle types

Electrification rates are expected to vary across different vehicle categories. Last-mile delivery and public transit fleets are likely to lead the electrification charge due to their operational advantages. These segments are well suited for electrification due to factors such as short distances, frequent stops, and the ability to return to central bases for charging. While transit and delivery vehicles may pioneer electrification, other segments like freight trucking hold promise. Despite their potential, electrification of heavy-duty vehicles faces challenges related to battery capacity and the required charging infrastructure (Figure 1.9) [32–35].

1.7 Preparation for extensive fleet electrification

The rapid acceleration of fleet electrification necessitates through preparation and planning. Utilities are already assisting fleet operators with infrastructure and advisory services, but the scale and complexity of this transformation necessitate grid upgrades and expansions to ensure reliability.

1.8 Anticipating long-term consequences

National Grid and Hitachi Energy conducted a comprehensive study to delve into the long-term consequences of large-scale fleet electrification. By studying real-world

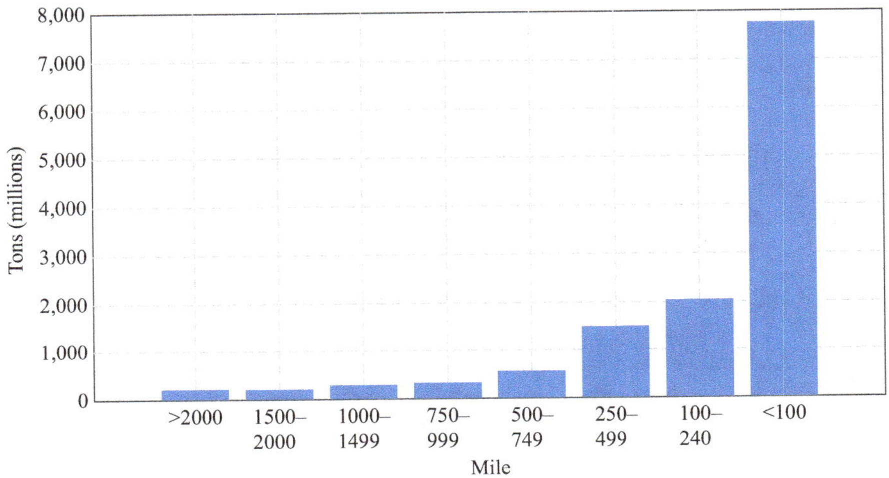

Figure 1.9 Truck freight (by weight) by distance traveled

fleets within National Grid's service area, this study explores the impacts of both fleet and light-duty vehicle electrification on infrastructure planning.

1.9 The favorable shift in total cost of ownership towards electric fleets

While EVs may present a higher initial cost when compared to traditional ICE vehicles, this initial expense can act as a hurdle to widespread adoption. However, the commercial fleet sector often bases decisions on the overall lifetime expenses, accounting for both upfront and operational costs. In this regard, EVs typically demonstrate lower operating expenses (fuel and maintenance) compared to gasoline or diesel vehicles, making a strong case for EVs from a total cost of ownership (TCO) perspective.

The concept of TCO is increasingly aligning with many EV models designed for fleet usage. For example, a 2018 assessment conducted by the North American Council for Freight Efficiency (NACFE) revealed that numerous medium-duty vehicles have already achieved cost parity, and heavy-duty vehicles are expected to follow suit in the near future. Similarly, according to BNEF's 2021 analysis, the TCO of all-electric vehicles is already lower than that of ICE vehicles in specific scenarios, especially within regional and long-haul operations.

1.10 The oncoming impact of fleet electrification on the electric grid

While the transition to electric fleets may seem gradual compared to the rapid adoption of passenger EVs, the rate of commercial EV adoption is expected to accelerate significantly. BNEF's projection for 2021 anticipates a staggering increase of over 150 times in the number of electric buses and commercial fleet vehicles between 2020 and 2025, surpassing the projected 4-fold increase in electric passenger vehicles. This impending acceleration holds the potential to generate substantial consequences for fleets, communities, and the supporting electric distribution and transmission infrastructure.

It's worth noting that even if policy and corporate objectives are set for targets a decade or more in the future, the impacts might manifest earlier as the adoption of EVs continues to expand.

1.11 Grid impact of fleet electrification

Fleet operators may embark on substantial procurements simultaneously, driven by the need to standardize fleet operations, benefit from procurement economies of scale, or due to lease agreements reaching their conclusion. Unlike the passenger vehicle segment, which tends to be influenced by upfront costs, fleet operators might expedite transitions once the TCO of EVs falls below that of ICE vehicles. This economic-driven transition could lead to significant changes in electric load dynamics at fleet sites, creating contrasts with the more gradual growth observed in residential and other commercial charging environments such as public and workplace charging.

Fleet operators may also necessitate larger electric vehicle supply equipment compared to public, work, or residential charging setups. Consequently, although the individual charging sessions might be shorter, the demand for electricity at any given moment could be higher. This scenario could rapidly transform individual fleet sites into megawatt-scale loads. In addition, MHDVs are expected to require more energy per charging session due to their larger battery sizes and increased vehicle miles traveled.

Considering the scale of charging requirements for MHDVs, they are poised to contribute significantly to the aggregate electricity demand stemming from EVs. Recent estimates from BNEF indicate that electric buses and trucks might necessitate only 1.3% of all available charging ports but could account for approximately 25% of the total electricity consumption attributed to electric transportation.

Utilities are already engaged in supporting fleet operators through initiatives like make-ready infrastructure, fleet advisory solutions, and customer support services. However, the comprehensive electrification of entire fleets introduces novel complexities. The scale and scope of such an endeavor pose unique challenges for utilities and system operators. It becomes essential to accurately quantify and pinpoint the potential impacts of fully electric fleets on the electric grid, facilitating the large-scale electrification of fleets while minimizing costs for both fleet operators and other electricity consumers. Utilities must proactively plan for transmission and distribution investments in the coming decade to ensure a robust and reliable electricity grid capable of accommodating the increased demand arising from the electrification of fleets (Table 1.2).

Table 1.2 Summary of impact analysis sensitivities

	Higher grid impact	**Lower grid impact**
Fleet operations	Rapid, high-rate charging is favored for operational needs, especially during peak periods (e.g., between Black Friday and Christmas for delivery companies or during local emergency situations). • Charging optimization is hindered by fixed vehicle schedules.	• Vehicles cover fewer daily miles on average. • Vehicles make use of on-route charging and minimize depot charging, thereby decreasing the concentrated loads on specific power feeders. • Vehicle dwell times are longer than initially estimated, sometimes extending to days or weeks of idle time. • Battery efficiency experiences a greater than expected increase.
Fleet location	• Newly discovered fleets that were not considered in the study transition to electric vehicles such as small and local businesses.	Fleets gradually move to more easily accessible areas of the electric network, a phenomenon

(Continues)

Table 1.2 (*Continued*)

	Higher grid impact	**Lower grid impact**
	• Fresh fleets relocate to restricted areas, possibly due to advantageous access to transportation infrastructure, or existing fleets grow in size.	sometimes referred to as "headroom hunting."
Fleet vehicle types and numbers	Fleet sizes exceed initial estimates. • The proportion of heavy-duty vehicles in fleets is higher than initially anticipated. • Fleets move vehicles to address seasonal demands such as holiday deliveries.	Fleet sizes are smaller than originally estimated. • The proportion of light-duty vehicles in fleets is greater than initially predicted.
Charging infrastructure	• Fleets exceed their initial expectations by installing more fast-charging ports or higher-capacity chargers, irrespective of the added cost. • Fleets increase the number of charging ports to accommodate simultaneous charging of more vehicles, resulting in a decreased EV-to-EVSE (electric vehicle to electric vehicle supply equipment) ratio.	• Fleets install a lower number of fast-charging ports than initially expected. • Fleets possess a reduced quantity of charging ports for facilitating simultaneous charging, thereby increasing the EV-to-EVSE ratio.
Temperature impact	Fleets increase the installation of charging ports to counterbalance the decrease in battery efficiencies observed at lower temperatures.	Fleets experience reduced charging time to achieve a full fleet charge due to improved battery efficiencies in higher temperatures. In addition, advancements in battery technology enhance the overall performance during colder weather conditions.
Other technologies	The implementation of electric heating solutions, like electric heat pumps, speeds up the process of electrification and introduces an extra load on the electric system.	• Hydrogen (or alternative fuels like natural gas or diesel) gains a greater portion of the market for medium- and heavy-duty vehicles (MHDVs). • The performance and cost competitiveness of energy storage and other distributed energy resources (DERs) enhanced. • Managed charging proves effective in significantly decreasing peak loads across various fleet types and usage scenarios.

1.11.1 Assessing the long-term effects of extensive fleet electrification

To thoroughly examine the long-term effects of extensive fleet electrification, National Grid and Hitachi Energy conducted a comprehensive study centered around real fleets within National Grid's US service area. The study sought answers to two primary questions:

1. **Impact of fully transitioning a specific fleet to electric vehicles**
   This study examined the consequences of converting an entire fleet to EVs, emphasizing a 100% conversion of fleet vehicles to EVs. It considered different vehicle types and their operational patterns.
2. **Implications of simultaneous electrification of multiple fleets**
   The study explored what happens when multiple fleets in a region switch to EVs concurrently. This analysis assessed the collective electricity demand impact and potential overloads on distribution feeders.

Key study highlights

The analysis focused on fleet electrification and did not include the added load from light-duty passenger vehicles (e.g., home and public charging). Nevertheless, it provided an overview of how electric passenger vehicles could strain substation capacity when combined with fleet electrification.

The study established a "bottom-up" case study framework, offering insights into the influence of large-scale fleet electrification on a specific segment of the electric grid. It laid the groundwork for future analyses in three key areas:

- Fine-tuning assumptions about fleet usage and timelines: Obtaining more detailed data on fleet sizes, locations, adoption schedules, and operational patterns is crucial to understanding specific grid impacts. Further analysis will be needed to scale these impacts with total fleet adoption and timing, facilitating optimal investment scheduling.
- Examining broader system requirements and solutions: A system-level approach, considering various impacts (e.g., light-duty vehicle electrification, heating), will help stakeholders understand and address broader effects. Comparing multiple solutions will identify the most effective ones.
- Identifying policy considerations: Opportunities to propose new processes, programs, and frameworks to meet customer and system needs should be explored.

The study's findings hold significance for a broad range of stakeholders committed to decarbonizing commercial transportation, including utilities operating at distribution and transmission levels, fleet operators, regulators, policymakers, communities, suppliers, and various other interested parties.

Distribution feeder impacts

Figure 1.10 illustrates the impact of fleet electrification on 19 feeders in the study area under a "full charging" strategy. Winter conditions, with lower EV battery efficiency and higher charging demands, accentuated the effects. Of the 19 feeders:

- Over 68% (13 feeders) would eventually experience overload or overload risk when adjacent fleets transition to EVs.
- Four feeders (8, 10, 16, and 18) would exceed their capacity in both summer and winter due to fleet electrification.
- Feeder 18 was significantly affected, electrifying over 400 fleet vehicles, including regional freight vehicles with 150 kW chargers.
- Feeder 16 faced challenges due to its lower feeder voltage (4.8 kV).

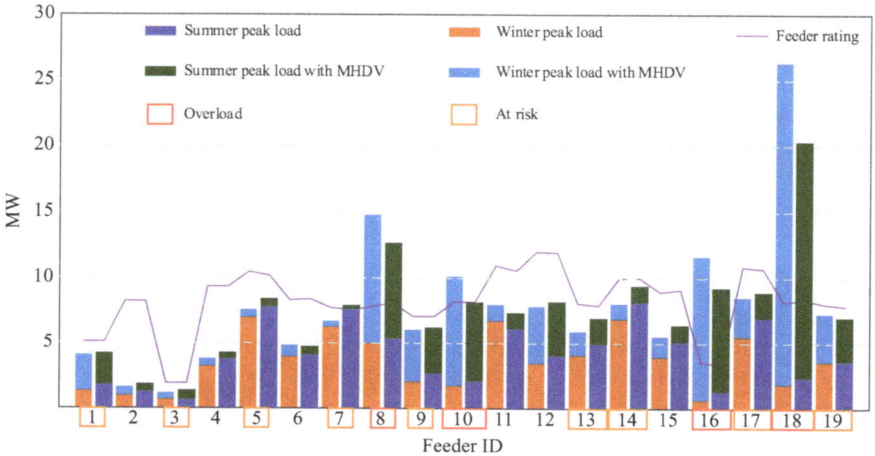

Figure 1.10 Impact of fleet electrification on the feeders during winter and summer peak load with full charging strategy

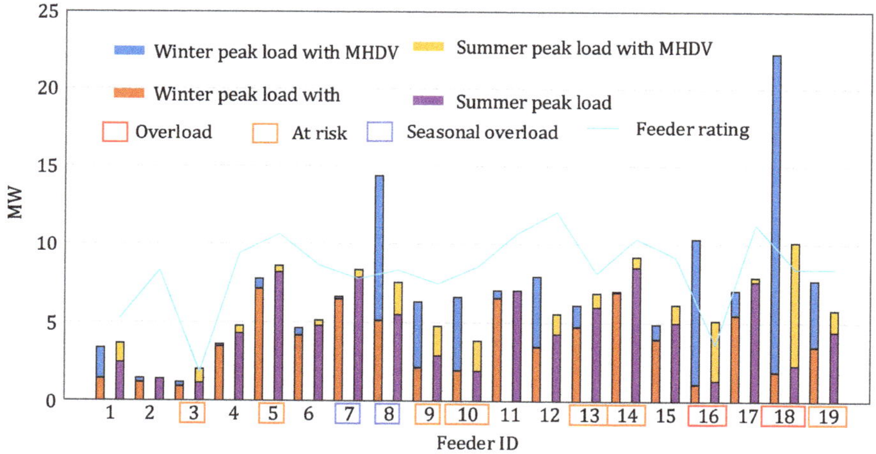

Figure 1.11 Impact of fleet electrification on the feeders during winter and summer peak load with minimum charging strategy

- Nine other feeders approached or exceeded their capacity during one season or reached 80% of capacity.

Minimum charging strategy

The minimum charging strategy, where fleet vehicles charge at the lowest rate to achieve a full charge before departure, was explored. It showed that Feeder 10 could handle fully electric fleets without overloading. Feeder 1 and Feeder 17 moved out of the risk category. However, 58% of feeders would still face overload or overload risk, with minor reductions during winter (Figure 1.11).

Substation and broader electrification impacts

The study considered concurrent electrification of light-duty vehicles alongside MHDVs, potentially amplifying the electrification's effects. To assess substation impacts:

- Figure 1.12 evaluated a specific substation serving seven feeders, including Feeder 17 and Feeder 1.18, which were heavily impacted by fleet electrification.

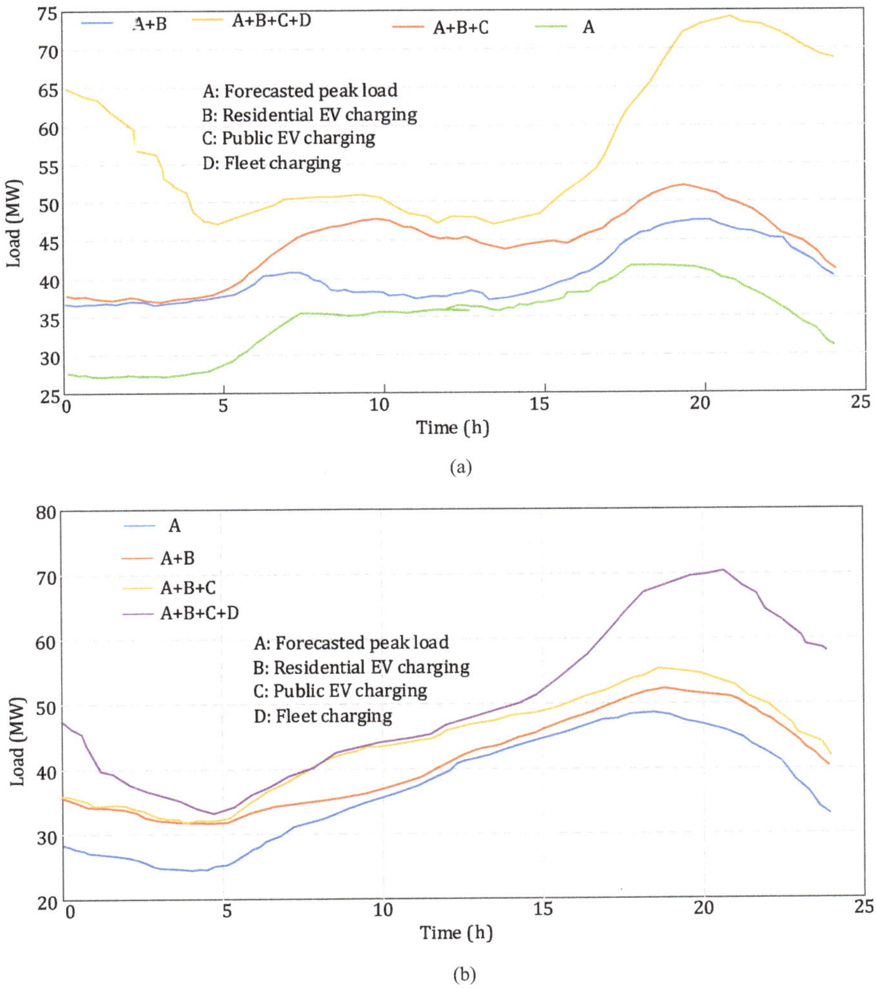

Figure 1.12 Impact of 100% fleet, residential and public charging at the substation: (a) winter-full charging strategy and (b) summer-full charging strategy

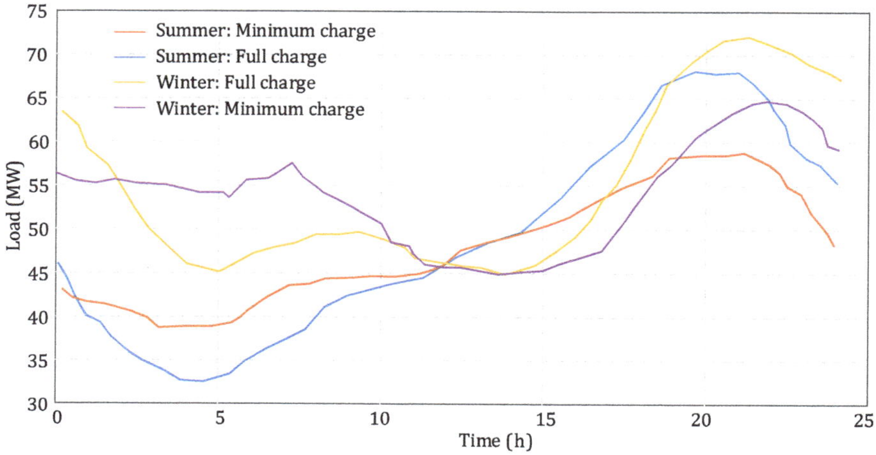

Figure 1.13 Impact of fleet charging strategy (full charging vs minimum charging) on substation loading; load includes residential, public, and fleet estimates

- The analysis considered peak loads during summer and winter, adding load from "100% electrification" scenarios for light-duty EV charging.
- The addition of residential and public charging increased substation demand by 20%, while electrified fleets raised it by over 60%.
- Some substation transformers could become overloaded despite adequate combined station capacity due to load imbalances.

Effectiveness of minimum charging strategy

Figure 1.13 compared the minimum charging strategy with the full charging strategy's impact on substation demand during summer and winter. It reduced peak load by approximately 10 MW in summer. Load shifting to morning hours in summer suggested an effective strategy. However, it's important to explore other options, such as energy storage, for addressing peak demand.

References

[1] Zhao J, Xi XI, Na QI, Wang S, Kadry SN, and Kumar PM (2021). The technological innovation of hybrid and plug-in electric vehicles for environment carbon pollution control. *Environ Impact Assess Rev*. 86:106506. https://doi.org/10.1016/j.eiar.2020.106506.

[2] Filote C, Felseghi R, Andreea R, Maria S, and Aşchilean I (2020). Environmental impact assessment of green energy systems for power supply of electric vehicle charging station. *Int J Energy Res*. 44:5678. https://doi.org/10.1002/er.5678.

[3] Johnson OD, and Hassan KA (2016). Issues of power quality in electrical systems. *Int J Energy Power Eng.* 5(4):148–154.

[4] Project "MERGE" – Mobile Energy Resources in Grids of Electricity, European project supported within the Seventh Framework Program for Research and Technological Development with Contract No.: 241399, http://www.evmerge.eu/ (accessed July 2014).

[5] https://www.iea.org/energy-system/transport/electric-vehicles.

[6] Zhou X, Zou L, Ma Y, and Gao Z (2017). Research on impacts of the electric vehicles charging and discharging on power grid, *2017 29th Chinese Control And Decision Conference (CCDC)*. 1398–1402, https://doi.org/10.1109/CCDC.2017.7978736.

[7] Lucas A, Bonavitacola F, Kotsakis E, and Fulli G (2015). Grid harmonic impact of multiple electric vehicle fast charging. *Electric Power Syst Res.* 127:13–21. https://doi.org/10.1016/j.epsr.2015.05.012.

[8] Delgado J, Faria R, Moura P, and de Almeida AT (2018). Impacts of plug-in electric vehicles in the Portuguese electrical grid. *Transp Res Part D: Transport Environ.* 62:372–385. https://doi.org/10.1016/j.trd. 2018.03.005.

[9] Williamson SS, Rathore AK, and Musavi F (2015). Industrial electronics for electric transportation: current state of-the-art and future challenges. *IEEE Trans Ind Electron.* 62:3021–3032.

[10] Grunditz EA, and Thiringer T (2016). Performance analysis of current BEVs based on a comprehensive review of specifications. *IEEE Trans Transp Electr.* 2:270–289.

[11] Alamir, PN, Abdel-Rahim O, Ismeil M, Orabi M, and Kennel R (2018). Fixed frequency predictive MPPT for phase-shift modulated LLC resonant micro-inverter, *2018 20th European Conference on Power Electronics and Applications (EPE'18 ECCE Europe)*, Riga, Latvia.

[12] Farhoodnea M, Mohamed A, Shareef H, and Zayandehroodi H (2013). Power quality impacts of high-penetration electric vehicle stations and renewable energy-based generators on power distribution systems. *Measurement.* 46(8): 2423–2434.

[13] World Resources Institute, How Did Shenzhen, China Build World's Largest Electric Bus Fleet? https://www.wri.org/blog/2018/04/how-did-shenzhen-china-build-world-s-largest-electric-bus-fleet (accessed 29 June 2019).

[14] Mindykowski J, and Tarasiuk T (2015). Problems of power quality in the wake of ship technology development. *Ocean Engineering.* 107:108–117. 10.1016/j.oceaneng.2015.07.036.

[15] Rehmatulla N, Calleya J, and Smith T (2017). The implementation of technical energy efficiency and CO_2 emission reduction measures in shipping. *Ocean Engineering.* 139:184–197. 10.1016/j.oceaneng.2017.04.029.

[16] Freire R, Delgado J, Santos JM, and de Almeida AT (2010). Integration of renewable energy generation with EV charging strategies to optimize grid load balancing, *13th International IEEE Conference on Intelligent Transportation Systems*, pp. 392–396, https://doi.org/10.1109/ITSC.2010.5625071/.

[17] Pillay P, and Manyage M (2001). Definitions of voltage unbalance. *IEEE Power Eng Rev*. 21(5):49–51. https://doi.org/10.1109/MPER.2001.4311362.
[18] Wu X, Aviquzzaman Md, and Lin Z (2015). Analysis of plug-in hybrid electric vehicles' utility factors using GPS-based longitudinal travel data. *Transport Res Part C: Emerging Technol*. 57:1–12. https://doi.org/10.1016/j.trc.2015.05.008.
[19] Clairand J-M, Guerra-Terán P, Serrano-Guerrero X, González- Rodríguez M, and Escrivá-Escrivá G (2019). Electric vehicles for public transportation in power systems: a review of methodologies. *Energies*. 12(16):3114.
[20] Chen H, Jia Y, Hu Z, Wu G, and Shen ZJM (2018). Data-driven planning of plug-in hybrid electric taxi charging stations in urban environments: a case in the central area of Beijing, *Proceedings of the 2017 IEEE PES Innovative Smart Grid Technologies Conference Europe, ISGT-Europe 2017*, vol. 2018, pp. 1–6.
[21] Tahir Y, Khan I, Rehman S, *et al.* (2021). A state of the art review topologies and control techniques of solid-state transformers for electric vehicle extreme fast charging. *IET Power Electronics*. 14(9):1560–1576.
[22] Khalid M, Rizwan A, Mohammad S, Sarwar A, and Jamil Asghar MS (2019). A Comprehensive review on electric vehicles charging infrastructures and their impacts on power-quality of the utility grid. *eTransportation*. 1:100006. https://doi.org/10.1016/j.etran.2019.100006.
[23] Habib S, Khan M, Abbas F, *et al.* (2020). Contemporary trends in power electronics converters for charging solutions of electric vehicles. *CSEE J. Power Energy Syst.* 6(4):911–929. https://doi.org/10.17775/CSEEJPES.2019.02700.
[24] Liang X (2017). Emerging power quality challenges due to integration of renewable energy sources. *IEEE Trans Ind Appl*. 53(2):855–866. https://doi.org/10.1109/TIA.2016.2626253.
[25] Hossain E, Tür MR, Padmanaban S, Ay S, and Khan I (2018). Analysis and mitigation of power quality issues in distributed generation systems using custom power devices. *IEEE Access*. 6:16816–16833. https://doi.org/10.1109/ACCESS.2018.2814981.
[26] Gandoman FH, Ahmadi A, Sharaf AM, *et al.* (2018). Review of FACTS technologies and applications for power quality in smart grids with renewable energy systems. *Renew Sustain Energy Rev*. 82(1):502–514. https://doi.org/10.1016/j.rser.2017.09.062.
[27] Heydt GT (1983). The impact of electric vehicle deployment on load management strategies. *IEEE Power Eng Rev*. PER-3(5):41–42.
[28] Rahman S, and Shrestha GB (1993). An investigation into the impact of electric vehicle load on the electric utility distribution system. *IEEE Trans Power Deliv*. 8(2):591–597.
[29] Sortomme E, Hindi M, MacPherson S, and Venkata S (2011). Coordinated charging of plug-in hybrid electric vehicles to minimize distribution system losses. *IEEE Trans Smart Grid*. 2(1):198–205.

[30] Morais H, Sousa T, Vale Z, and Faria, P (2014). Evaluation of the electric vehicle impact in the power demand curve in a smart grid environment. *Energy Convers Manag.* 82:268–282.
[31] Abdel-Rahim O, Orabi M, and Ahmed ME (2010). Development an efficient photovoltaic (PV) configuration for low power applications, 2010 IEEE International Conference on Power and Energy, Kuala Lumpur, Malaysia, pp. 622–627.
[32] Leemput N, Geth F, Van Roy J, Büscher J, and Driesen J (2015). Reactive power support in residential LV distribution grids through electric vehicle charging. *Sustain Energy Grids Netw.* 3: 24–35.
[33] Sousa T, Morais H, Vale Z, and Castro R, (2015). A multi-objective optimization of the active and reactive resource scheduling at a distribution level in a smart grid context. *Energy.* 85: 236–250.
[34] Steinberg D, Bielen D, Eichman J, *et al.* (2017). *Electrification and Decarbonization: Exploring U.S. Energy Use and Greenhouse Gas Emissions in Scenarios with Widespread Electrification and Power Sector Decarbonization*, National Renewable Energy Laboratory. Technical Report NREL/TP-6A20-68214.
[35] Panossian N, Muratori M, Palmintier B, and Meint A (2022). Challenges and opportunities of integrating electric vehicles in electricity distribution systems. *Curr. Sustainable Renewable Energy Rep.* 9: 27–40.

Chapter 2

Regulation and standardization of transportation electrification: components, specifications, safety, and communications

Abdelmonem Draz[1], Ahmed M. Othman[1] and Attia A. El-Fergany[1]

Recently, the reliance of electric networks on Distributed Generation in terms of Renewable Energy Resources has become imperative instead of conventional fossil fuel sources. In this context, transportation electrification approach attracts the attention of stakeholders for replacing gasoline vehicles with electric vehicles (EVs). Not only the electrification terminology is emitted in private cars scale but also in the scale of electric buses, electric trains, and electric ships. For this reason, releasing well-established international standards becomes mandatory to unify design procedure, technical requirements, and components specifications in the transportation electrification market. As a result, this chapter offers a broad discussion of the many norms and laws governing the transportation electrification industry, whether in communities in Europe or America. Additionally, several EV kinds are discussed, including fuel cell EVs, plug-in EVs, and hybrid EVs. In addition, standardization of waterborne transportation systems with the transition from fossil fuels into hydrogen-based fuels and batteries is also investigated.

Electric charging can be categorized into two techniques: conductive charging and wireless charging. The first technique is called when cable and connector are used while the second one is named where there is not any hardware connection between the EV and charging facility. In this regard, inductive power transfer is deemed as one of the most technologies in wireless charging infrastructure using two coils: transmitter coil and receiver coil. AC current passes through the transmitter coil which in turn produces AC magnetic flux that results in induced voltage in the receiver coil causing battery charging. On the other side, conductive charging can be classified into slow, fast, or ultra-fast charging based on the charging power and time.

[1]Electrical Power and Machines Department, Faculty of Engineering, Zagazig University, Zagazig, Egypt

Normally, charging with AC power is treated as slow charging while DC charging is the solution of fast charging. It is worth mentioning that recommended charging voltages and currents have been stipulated in the international standards. Technical requirements of connectors, plugs, and socket-outlets are indicated in the standards in addition to manuals, catalogues, and datasheets provided by the manufacturers in the transportation electrification market.

In this chapter, multiple scales of communication topologies in addition to several charging control techniques are also investigated. Communication interfaces between EVs and the charging facility may be formed in the basic level or in high intelligent level through human–machine interface. Afterward, various control techniques either in charging or discharging are also examined through the battery management system. Methods of communication between EVs and the distribution grid are also highlighted for the purpose of intensifying knowledge. The most common interactive sessions that regulate the power flow between EVs and the grid are thought to be vehicle to grid (V2G), grid to vehicle (G2V), vehicle to vehicle (V2V), and vehicle to X (V2X). Also, attempts are dedicated to clarify and scrutinize the regulations of smart charging infrastructure via local controller or smart centralized controller through specified cloud solution. Eventually, this chapter outlines the most techno-commercial relevant standards in the field of transportation electrifications either in manufacturing, design, or implementation stages.

2.1 Introduction

Indeed, the industry of electric vehicles (EVs) has become more popular nowadays because of its environmentally friendly behavior compared to internal combustion engine (ICE) vehicles [1,2]. However, many improvements are still dedicated to the design of EVs regarding the rechargeable batteries, driving motors, and controllers [3,4]. On the other hand, ICE vehicles are characterized by the long driving range compared to EVs in addition to the ease accessibility to the refueling stations [5,6].

Generally, EVs are vehicles which are powered by electric sources either from a battery, charging solar cell, or by fuel from generator or fuel cell [7,8]. Accordingly, EVs include various forms such as road vehicles like electric buses (EBs), rail vehicles or electric trains (ETs), underwater vessels, and electric aircraft [9,10]. Therefore, releasing well-established standards and regulations regarding design, installation, and safety of EVs components becomes mandatory [11,12]. Moreover, technical requirements of connectors and plugs shall be identified to unify the design process in the electrification market [13,14]. Communication, safety, and electromagnetic compatibility (EMC) requirements are also recognized in this chapter to facilitate the selection criterion [15,16].

2.2 Transportation electrification

2.2.1 Electric vehicles, buses, and trains

Figure 2.1 shows the typical configuration of modern EVs comprising various subsystems. The electric propulsion subsystem contains the driving electric motor with the required power converter while the energy source subsystem includes the battery and its charger. In this context, some modern EVs comprise a three-phase induction motor with the corresponding PWM inverter as the electric propulsion. Moreover, they may contain lead acid or nickel-based batteries in the energy subsystem along with an energy management unit. The auxiliary power supply inside the EV provides the necessary power with various voltage levels to the EV auxiliaries such as temperature control and power steering units [17].

2.2.2 Types and technologies of EVs

2.2.2.1 Battery EVs

They are called plug-in EVs (PEVs) that consist of storage battery, electric motor, and controller. They are plugged-in into the utility to charge the battery supplying the driving motor either in forward or reverse locomotion [18]. The market's attention is now headed to PEVs rather than conventional ICE vehicles due to some benefits concluded as follows:

- Low operating costs
- Low emissions
- Flexible fueling
- Energy security

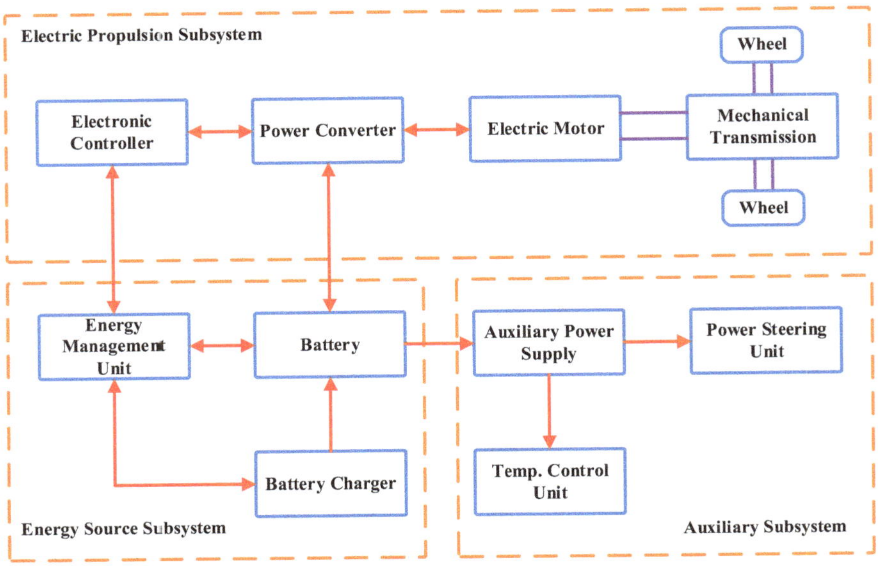

Figure 2.1 EV typical architecture

In this context, PEVs have much smaller operating costs and CO_2 emissions compared to ICE vehicles because of their dependency on clean and free renewable sources. Moreover, their refueling is very resilient as they can be charged at home, from another EVs, or from public charging stations. Since PEVs are less dependent on petroleum resources than in the case of ICE vehicles, the energy sector becomes more secure as the reliance on imported fuels is diminished.

2.2.2.2 Hybrid EVs

Hybrid EVs (HEVs) or partial zero emission vehicles (PZEVs) are combination of storage batteries, ICE or any electric propulsion source, regenerative braking, and driving motor. They are petroleum-based fuel vehicles that are charged through ICE or the propulsion source and are not plugged into the charger. HEVs can be categorized into two main configurations: series and parallel configurations. In the series arrangement, HEVs are driven by electric motors which are supplied by storage batteries or ICE or both. Contrarily, parallel HEVs are driven by electric motors or by ICE that work directly to the driving wheels via the transmission system. It is worth mentioning that in both arrangements, regenerative braking can be employed converting the motor into generator unit for charging the battery [17,18].

Outspokenly, the parallel HEVs configuration is much cheaper and smaller compared to the series arrangement which makes it a widely used solution. Conversely, series type application is limited due to its high cost which makes it a candidate solution in electric ships. Moreover, they are distinct by transmission of electric energy through generator-motor set besides a separate control for each motor wheel. Hereinafter, the parallel HEVs arrangement can be represented using the degree of hybridization factor (DHF) as defined in (2.1).

$$DHF = \frac{P_{motor}}{P_{motor} + P_{ICE}} \tag{2.1}$$

where P_{motor} denotes the electric motor power, and P_{ICE} denotes the ICE power. Consequently, the greater DHF value, the larger possibility of using smaller ICE in which its operation is almost near the optimum efficiency.

2.2.2.3 Plug-in hybrid EVs (PHEVs)

They involve the features of PEVs and HEVs in which the battery is charged via plugging into the charger or through ICE or the propulsion source. It is worth noting that PHEVs have a longer driving range compared to HEVs as they contain larger battery packs. In addition, most PHEVs utilize the power of ICE during rapid acceleration or even in severe heating or cooling demands.

2.2.2.4 Fueled EVs (FEVs)

FEVs employ fuel-based sources such as fuel cell or metal/air battery instead of rechargeable batteries as in the case of battery EVs. These types of EVs require hydrogen fuel tanks to be stored where this technology is hard to implement. Therefore, they are not popular in private cars that deploy on-board chargers. However, they can be considered as the preferable solution in public service vehicles such as EBs and ETs as they are charged at only one common place. Furthermore, metal/air batteries are exploited in this technology as an analogous to fuel cell which are refueled by replacing the metal electrodes. In this regard, zinc/air (Zn/Air) batteries are deemed as the superior type in this class which results in various commercial Zn/Air EBs in this decade [18].

2.2.3 Electric waterborne transportation

Since the world's attempts are dedicated to a clean environment, the maritime industry is electrified for transportation of passengers and goods. Accordingly, the transition of waterborne mobility becomes mandatory for a more efficient and sustainable environment. In this context, hydrogen-based fuel such as fuel cells and batteries are incorporated as alternative sources in waterborne transportation industry. It is worth mentioning that this industry can be included in public transportation systems which operate on scheduled timetables. Waterborne mobility can be considered as a socio-technical system from economic, social, and environmental perspectives. Moreover, this industry is a multi-level perspective and passes through three levels: socio-technical regime, landscape, and niche level [19].

Stakeholders model is employed for further analysis of waterborne transportation system by incorporating the Mendelow's matrix which is divided into four regions. These four squares are dependent on the stakeholder's power and interest to attain specific levels of satisfaction and priority. Furthermore, some countries, like Sweden and the United States, developed a list of regulations to support the future of maritime industry from financial, technical, and environmental perspectives. The proposed regulations set up recommendations to lessen the use of fossil fuels in the waterborne sector and hence reduce the greenhouse gas emissions.

2.2.4 Electric vessels standardization

Free waterborne transportation supports the implementation of 100% free electric drive ferries promoting energy efficient and sustainable communities. However, E-ferries may require a charging power of 4 MW which represents a challenge to the designers. Moreover, there are technical and operational barriers in the E-ferry industry regarding the sailing range besides the high charging power. Also, ferries electrification experience heavy weights because of the large battery packs installed inside the E-ferry. Due to the increased fuel prices, the E-ferry industry is still booming in Europe utilizing the retrofitting concept [19].

2.3 International standards and regulations

2.3.1 Components specifications inside EVs

In this subsection, the technical requirements of batteries and electric motors are interrogated. Moreover, the integration between propulsion and energy source subsystems is highlighted focusing on pros and cons of each type [20,21].

2.3.1.1 Technical requirements of batteries

Voltage

Most EV cells are rated at 6 V or 12 V; therefore, they shall be connected in series to attain the required battery voltage for the EV. Consequently, the battery voltage decreases when supplying power to the electric motor and increases during charging process. This behavior as explained in (2.2) is due to the battery internal resistance (R) which can also be visualized through Figure 2.2.

$$V = E - IR \tag{2.2}$$

where E denotes the battery open circuit voltage and V is the load terminal voltage. In fact, the value of E is not constant and varies with temperature and battery state of charge (SoC).

Ampere-hour capacity

Normally, the battery capacity is rated in Ampere-Hour which defines the value of the supplied current in a specified discharging time. However, this value is not constant and varies according to the discharge rate. In this regard, let's say there is a battery with $C_5 = 100$; this means that this battery is rated at 100 AH for 5 hours. If the discharge time rises to 20 hours, the capacity will slightly increase. On the other side, if the fast discharge rate is employed, let's say for 1 hour, the capacity will fall below the C-rated value.

Specific energy

Battery specific energy is measured in Wh/kg that defines the amount of energy stored inside the battery packs per unit mass. On the other hand, specific power can define the battery performance which denotes the amount of delivered power per unit mass measured in $\frac{W}{kg}$. However, it is highly dependent on the load characteristics rather than the battery itself. Therefore, some batteries may have a very good specific energy, although they have a poor specific power. Accordingly, Ragone plot is a well-supportive figure in comparing the performance of batteries which correlates the specific power with specific energy.

Ampere hour efficiency

It is also called the charging efficiency which indeed is below than 100% and relies on the battery technology, temperature, rate of charge, and SoC. Practically, this

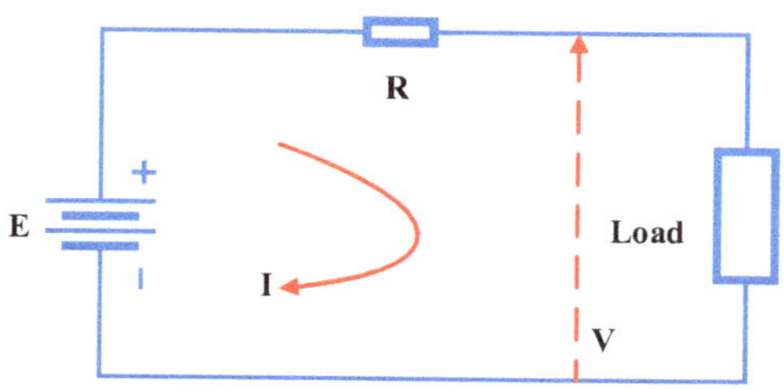

Figure 2.2 Battery simplified equivalent circuit

value is very close to 100% if the battery is charged from 20% to 80%; however, it falls down greatly if it is charged below 20%.

Energy efficiency

Batteries with high energy efficiency values are recommended in EV applications. Energy efficiency is defined as the ratio between the delivered electrical energy from the battery to the required electrical energy for restoring the battery to its initial state. It is worth mentioning that its value depends on the battery performance in charging or discharging processes. It decreases rapidly if the battery is subjected to high charging or discharging rates.

Self-discharge rate

It varies with temperature and battery type which indicates the rate at which the batteries discharge when left unused. The rate of self-discharge increases with high temperature which brings out avoiding keeping the batteries for a long time without recharging.

Battery life

The number of battery deep cycles governs the lifetime throughout of the battery that controls the EV running costs. Its value depends on the battery type, design details, and how the battery is used.

Hereinafter, the most widely used commercial EV batteries will be investigated which are summarized in Figure 2.3. A thorough talk is also highlighted for these commercial EV batteries with respect to their chemical reactions, merits, demerits, and applications.

Lead acid batteries

These batteries are the popular type used in EV applications due to their capability of withstanding deep cycle discharging in addition to their low internal resistance value. In this regard, the chemical reaction of lead acid batteries is shown in (2.3) where negative and positive plates are immersed in sulphuric electrolyte. In addition, as a rule of thumb, (2.4) is considered as a good estimator in computing the internal resistance value [20].

$$Pb + PbO_2 + 2H_2SO_4 \leftrightarrow 2PbSO_4 + 2H_2O \tag{2.3}$$

$$R = N_{cells} \times \frac{0.022}{C_{10}} \tag{2.4}$$

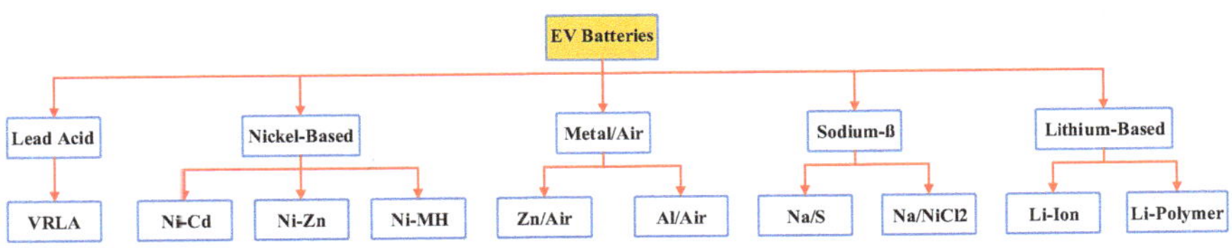

Figure 2.3 EV commercial batteries

where N_{cells} is the number of series connected cells, and C_{10} denotes the ampere hour capacity at 10-hour discharge time.

As a modern type of these batteries, valve-regulated lead acid (VRLA) batteries are utilized to boost the safety precautions. This can be done using a valve to release the gas at a specific pressure besides using a gel electrolyte instead of a liquid one. It is necessary to keep VRLA batteries in a good *SoC* level to avoid the sulphation problem where an insulating layer is formed over the battery electrodes. Furthermore, multiple step charging technique shall be followed in the charging of VRLA batteries as per most of manufacture's recommendations. In this technology, the cell is charged until it reaches a predetermined value; then the current is switched off until it decays to a certain point; afterward, the current is switched on again.

Nickel-based batteries

Many commercial batteries use nickel as the positive electrode such as nickel cadmium (Ni-Cd), nickel zinc (Ni-Zn), and nickel metal hybrid (Ni-MH) batteries. However, Ni-Zn batteries are rarely used due to their limited life cycle of about 300 deep discharging cycles. On the other side, Ni-Cd batteries represent the main competitor to VRLA batteries in EV applications as they have almost double the specific energy. Their formula of chemical reaction is presented in (2.5), while their internal resistance can be calculated from (2.6). Although Ni-Cd batteries have a small internal resistance value, they are slightly larger than lead acid batteries as the voltage of Ni-Cd cell is about 1.2 V while it is 2 V in lead acid type. Consequently, more series cells are required in Ni-Cd compared to VRLA batteries although they are less prone to self-discharge phenomenon [21].

$$Cd + 2N_iOOH + 2H_2O \leftrightarrow Cd(OH)_2 + 2N_i(OH)_2 \tag{2.5}$$

$$R = N_{cells} \times \frac{0.06}{C_3} \tag{2.6}$$

Ni-MH batteries are similar to Ni-Cd type except for that the used negative electrode is the hydrogen which is imitated as fuel cell. These types were introduced commercially in the last decade as they have larger energy and power density compared to Ni-Cd type in addition to their fast-charging rates.

Metal/air batteries

The charging concept of metal/air batteries is different as they cannot be recharged by reversing the current but by replacing the metal electrodes with new ones. They are classified into two main types: zinc/air (Zn/Air) and aluminum/air (Al/Air). The latter type has a critical drawback of low specific power, while Zn/Air batteries have better performance and characteristics.

Sodium-based batteries

These types of batteries use a liquid sodium negative electrode which is characterized by operating at high temperatures. Moreover, one or more liquid electrodes can be implemented forming a molten sodium enabling them suitable for heavy duty applications. They are classified into two main categories: sodium/

sulphur (Na/S), and sodium metal chloride ($Na/NiCl_2$) or Zebra batteries. Na/S cells are featured by high specific energy which is almost six times of lead acid batteries, while Zebra cells have higher safety level as they have solid positive electrolyte separated from the molten sodium.

Lithium-based batteries

These batteries offer the highest energy density among all types although they are expensive technology. They are categorized into lithium/ion (Li-Ion) and lithium/polymer batteries. Li-ion cells offer two types of electrolytes, either liquid or solid polymer. In addition, controlling the charging voltage is a very big issue because if its rate is so high, it will damage the battery. Contrarily, if it is so low, the battery will not be sufficiently charged. Eventually, Table 2.1 lists a technical comparison between VRLA, Ni-Cd, and Ni-MH batteries.

2.3.1.2 Specifications of rotating electric machines

Proper selection and sizing of electric motors inside the EVs are very crucial to drive the EV easily and efficiently. Therefore, a brief talk with technical comparisons between commutator and brushless motors is analyzed. Furthermore, Figure 2.4 presents the most popular used motors for transportation electrification systems [22].

Table 2.1 Technical requirements of most popular commercial EV batteries

	VRLA	Ni–Cd	Ni–MH
Specific energy (Wh/kg)	20–35	40–55	≈65
Specific power (W/kg)	≈250	≈125	200
Energy density (Wh/L)	54–95	70–90	≈150
Nominal cell voltage (V)	2	1.2	1.2
Ampere hour efficiency	≈80%	Good	Quite good
Internal resistance (Ω)	0.022 for 1 AH	0.06 for 1 AH	0.06 for 1 AH
Operating temperature (°C)	Ambient	−40 to +80	Ambient
Self-discharge rate	≈2% per day	0.5% per day	5% per day
Recharging time (h)	8	1	1
Number of life cycles (to 80% capacity)	Up to 800	Up to 1,200	Up to 1,000

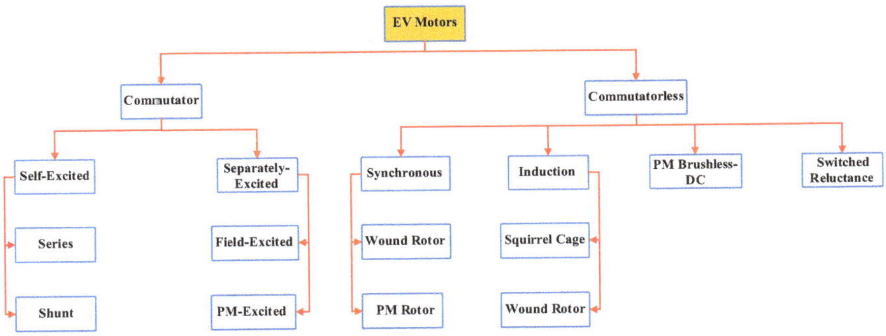

Figure 2.4 EV commercial electric motors

Commutator brushed DC electric motor
Controlling these types of motors is an easy task by adjusting the applied voltage which in turn controls motor speed and torque. In another way, flux control technique may be implemented resulting in a wide range of motor speed and torque that is popular in EV applications. The latter technique is employed in the case of using electromagnetic coils or field winding placed on the stator rather than using permanent magnets. There are three main categories of this technology which include series, shunt, and separately excited DC motors.

Brushless electric motor
The motivation to use brushless motors is due to the high friction losses between brushes and commutator occurring in commutator DC motors. Brushless DC, switched reluctance, and induction motors represent the most widely used types for transportation electrification. Obviously, brushless DC motor is an effective solution in high specific power requirements due to installing permanent magnet in the rotor circuit. On the other side, the rotor of switched reluctance motor is not a permanent magnet which makes it a preferable option in high-speed applications. However, induction motors are the most widely used technology in EV applications due to their simple operation and spectacular performance. In other words, the complexity of the produced AC power from inverters for induction motors is less than its counterpart in brushless or switched reluctance motors.

2.3.2 Safety requirements in transportation electrification

Safety guidelines of EVs components are stipulated in the international standards and regulations for both manufacturers and operators [23]. In addition, they are committed to the stiffness and quality of the applicable materials besides the consistent performance over the lifetime. These safety requirements can be concluded as follows [24]:

- Technical requirements of EV components such as ingress protection, insulation resistance, and utilization category shall be compatible with the standards.
- Disconnectors or emergency switches shall be incorporated with the EV supply equipment (EVSE) to disengage it from the supply during maintenance actions.
- Every EV charger must be fitted with a separate protection component.
- Moreover, it shall be protected with an over current protector like circuit breaker or fuse with the aid of residual current protection for earth leakage protection.
- The rating of charging cable shall be suitable with the rating of the EVSE.
- The average length of the charging cable is about 5 or 7.5 m. However, it may be increased if it is fitted with a cable management system.
- Finally, the cable insulation category shall be corrosion-resistant inside the range of the operating temperature according to the EVSE class.

2.3.3 Charging techniques in transportation electrification

- *Charging modes as per IEC 61851*
  EVSE or EV charger are classified based on their power rating, voltage level, charging speed, and application. Therefore, IEC 61851 lists four standardized

Table 2.2 IEC 61851 standard charging modes

Mode type	Mode 1	Mode 2	Mode 3	Mode 4
Voltage	250 V (one-phase) 480 V (three-phase)	250 V (one-phase) 480 V (three-phase)	250 V (one-phase) 480 V (three-phase)	Up to 1,000 V DC
Current (A)	16 A	32 A	250 A	400 A
Power (kW)	3.7 kW	22 kW	50 kW	≥50 kW

Table 2.3 North American standard charging levels

Charging	Level 1	Level 2	Level 3	Level 4
Type	AC	AC	DC	DC
Voltage (V)	120 V	240 V	500 V	800–1,000 V
Power (kW)	1.4 kW	3–20 kW	50 kW	≥400 kW
Time	43 h	11 h	60 min	20 min

charging modes which are concluded in Table 2.2. In mode 1 charging, the EV is connected to an AC supply with a socket-outlet of a maximum voltage of 250 V, one-phase or 480 V, three-phase with a maximum current of 16 A. In mode 2 charging, the charging current exceeds 32 A with the same voltage levels. In mode 3, the charging power and current are also increased, while in mode 4, the charging is performed using DC power for fast charging applications [25,26].

- *Charging levels as per North American standards*
  As per North American standards, EV charging can be classified into four levels according to charging: voltage, power, time, and connector type. Level 1 and level 2 charging are AC powered modes with 120 V and 240 V, respectively, with the charging power levels as indicated in Table 2.3. In level 3, DC power of about 50 kW is used for charging, while the power is increased to about 400 kW in level 4. It is worth noting that level 4 is employed for fast charging in public transportation charging infrastructures. In fact, CHAdeMO connector is utilized for DC charging, while J1772 connector is employed for AC charging [27].

2.3.3.1 Charging and discharging characteristics

The discharging process is characterized by voltage decaying resulting from the conversion of the stored chemical energy to beneficial electrical energy. Furthermore, the battery operating temperature will affect the discharging characteristics regarding the capacity and cell voltage. Accordingly, the thermal management system via temperature control unit is crucial to be installed inside the EV [28].

Whenever the battery reaches its cut-off voltage, they shall be recharged where their characteristics vary with charging current, time, and temperature. It is highly recommended to charge the battery until it reaches its full capacity to avoid overcharging and then overheating problems. In this context, three charging techniques can be followed: controlled current charging, controlled voltage charging, or both. Outspokenly, the charging method varies with the charging characteristics of the EV battery for efficient and safe operation. Therefore, methods such as low-rate constant current or high-rate pulsating current are not recommended in all cases.

- *Charge equalization*
  As the battery consists of some cells in series, it may be prone to that the current drawn from each individual cell is not equal. Accordingly, the *SoC* level of each cell is not equalized because of the variations in self-discharge rates due to manufacturing tolerances and temperature. In other words, cells with high self-discharge rates will not be fully charged other than the remaining cells. As previously mentioned, charging equalization will prohibit this phenomenon by fully charging the battery not near to full case. Indeed, some cells may be prone to overcharging issues; however, they must be within the accepted limits [29].

2.3.3.2 On-board vs off-board chargers

There are mainly two types of battery chargers called on-board and off-board chargers. The first type is usually installed inside the EV and shall be small size and weight and dedicated to slow and low-power charging. On the other hand, off-board chargers are installed outside EVs for fast and high-power charging. In this context, a simple architecture for on-board charger using AC power is shown in Figure 2.5 while the rectification and charging stages are executed inside the EV. Conversely, Figure 2.6 shows a simple module of off-board charger using DC power along with implementing a high-frequency transformer for galvanic isolation purposes. However, the galvanic isolation for on-board charger is achieved by the line frequency transformer installed before the rectification stage [30].

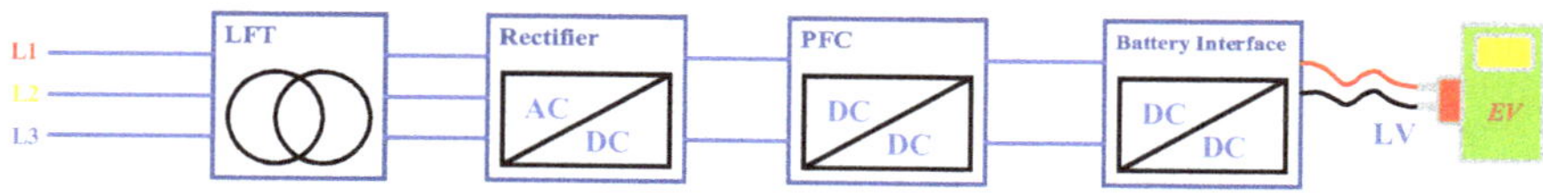

Figure 2.5 On-board charger architecture

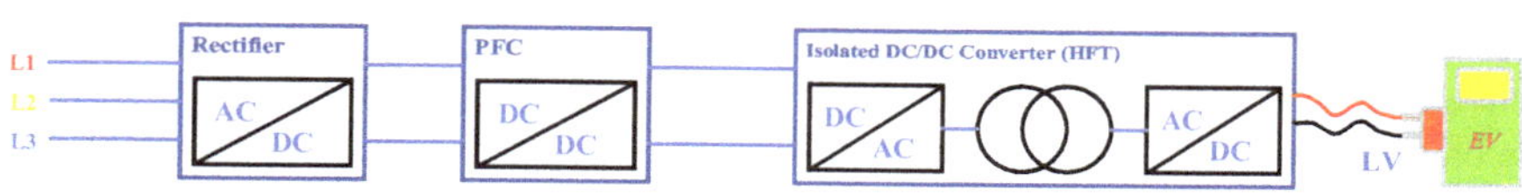

Figure 2.6 Off-board charger architecture

2.3.3.3 Technical requirements of connectors and plugs

As per IEC 62196, there are several conductive electrical interfaces between the EVSE and the EV that are summarized into universal, basic, DC, and combined interfaces [31]. In universal interface, there are 13 positions used in power or communication signals as indicated in Table 2.4. Therefore, the universal EV inlet shall be compatible with high-power AC or DC connectors. But for the basic interface, it contains seven position contacts that can be used in one-phase or three-phase connections as shown in Table 2.5. Subsequently, the one-phase connector is rated at 250 V, 32 A, while the three-phase coupler is rated at 480 V, 32 A [32,33].

Specifications of DC connectors are listed in Table 2.6 where they contain 12 possible contacts applicable for EV charging configuration AA and BB. Moreover, the PE contact is necessary to be installed in case of isolated or off-board chargers. In addition, Table 2.7 lists the technical requirements of combined connectors that can be used for AC or DC charging with nine possible contacts for the plug [34,35].

Table 2.4 Universal EV interface requirements

Position	High-power AC/AC	High-power DC/AC	Function
1	500 V, 250 A	600 V, 400 A	High-power AC or DC
2	500 V, 250 A	600 V, 400 A	High-power AC or DC
3	500 V, 250 A	–	High-power AC
4	480 V, 32 A	480 V, 32 A	L1 (phase one)
5	480 V, 32 A	480 V, 32 A	L2 (phase two)
6	480 V, 32 A	480 V, 32 A	L3 (phase three)
7	480 V, 32 A	480 V, 32 A	N (neutral)
8	Fault rated	Fault rated	PE (protective earth)
9	30 V, 2 A	30 V, 2 A	Control pilot
10	30 V, 2 A	30 V, 2 A	(+) Communication 1
11	30 V, 2 A	30 V, 2 A	(−) Communication 2
12	30 V, 2 A	30 V, 2 A	Data earth
13	30 V, 2 A	30 V, 2 A	Proximity

Table 2.5 Basic EV interface requirements

Position	AC		Function
	One-phase	Three-phase	
1	250 V, 32 A	480 V, 32 A	L1 (phase one)
2	–	480 V, 32 A	L2 (phase two)
3	–	480 V, 32 A	L3 (phase three)
4	250 V, 32 A	480 V, 32 A	N (neutral)
5	Fault rated		PE (protective earth)
6	30 V, 2 A		Control pilot
7	30 V, 2 A		proximity

Table 2.6 DC EV interface requirements

Position	Configuration		Function
	AA	BB	
1	600 V, 200 A	750 V, 250 A	DC +
2	600 V, 200 A	750 V, 250 A	DC −
3	30 V, 10 A	30 V, 2 A	Control pilot 1
4	30 V, 10 A	30 V, 2 A	Control pilot 2
5	30 V, 10 A	–	Control pilot 3
6	30 V, 2 A	30 V, 2 A	(+) Communication 1
7	30 V, 2 A	30 V, 2 A	(−) Communication 2
8	30 V, 2 A	–	Isolation monitor
9	–	750 V, fault rated	PE
10	30 V, 2 A	–	Proximity detection
11	–	30 V, 20 A	(+) Auxiliary power supply 1
12	–	30 V, 20 A	(−) Auxiliary power supply 2

Table 2.7 Combined AC/DC EV interface requirements

Position	Configuration		Function
	EE	FF	
1	600 V, 200 A	1,000 V, 200 A	DC +
2	600 V, 200 A	1,000 V, 200 A	DC −
3	–	–	DC −
4	–	–	DC −
5	–	–	DC +
6	–	–	DC +
7	600 V	1,000 V	PE
8	30 V, 2 A	30 V, 2 A	Control pilot
9	30 V, 2 A	30 V, 2 A	Proximity detection

2.3.3.4 EMC requirements in transportation electrification

As per IEC 61851-21 part 1 and 2, EMC requirements for EVs are stipulated. Generally, and during testing, EV shall operate within a voltage of +10% to −15% of the designed supply voltage and with ±1% variations in the supply frequency. Moreover, the engine of the EV under test shall be off, while it is in the charging mode. Tests shall be performed in an ambient temperature of 23°C± 5°C in a draught-free location. In addition, tests shall be carried out using the specified charging cable by the EVSE manufacturer [36].

Immunity tests shall be performed separately in a consecutive order where the battery *SoC* shall be maintained between 20% and 80% if the testing current cannot be regulated. However, if the current can be adjusted, it shall be kept at least at 80% of its rated nominal value. Meanwhile, harmonic currents shall be measured during testing up to the 40th order with the specified accepted limits as shown in Table 2.8.

Table 2.8 Maximum allowable harmonic currents

Harmonic order		Maximum harmonic current (A)
	Odd harmonics	
3		2.3
5		1.14
7		0.77
9		0.4
11		0.33
13		0.21
$15 \leq n \leq 39$		$0.15 \times 15/n$
	Even harmonics	
2		1.08
4		0.43
6		0.3
$8 \leq n \leq 40$		$0.23 \times 8/n$

Furthermore, the EV charging system shall have immunity to the radiated electromagnetic waves according to ISO 11451 part 2. In this context, frequency measurements shall be ranged between 20 and 2,000 MHz considering the middle of the EV is the reference point [36].

2.3.3.5 Conductive vs inductive charging

According to the energy transfer technology between the EVSE or the charger and the EV, the charging may be conductive or inductive. In conductive charging, a plug is used to feed power to the EV via socket and connector. On the other hand, inductive charging utilizes the principles of magnetic coupling between the EVSE and EV. It relies on the principle of induction between two coils of a high-frequency transformer, one installed in the charger and the other in the EV. Initially, the AC power with normal frequency (50 or 60 Hz) is rectified and converted to a high-frequency power (80–300 kHz) inside the charger [37]. Afterward, this AC power is transmitted to the EV via induction and then rectified to charge the EV battery as implemented in Figure 2.7.

One of the most widespread technologies of inductive charging is the inductive wireless power transfer technique. In this technology, an AC current passes through the transmitter coil producing a magnetic flux which in turn induces a voltage in the receiver coil. This induced voltage can charge the EV battery safely and efficiently. However, this technology is less efficient compared to conductive charging besides having a small power density.

2.3.4 Communication interfaces between EVs

In this subsection, deploying the smart charging schemes via different communication protocols is discussed. In this context, typical energy management system (EMS) configuration of EV architecture is implemented in Figure 2.8. As it is shown, various subsystems are incorporated such as sensors of battery voltage,

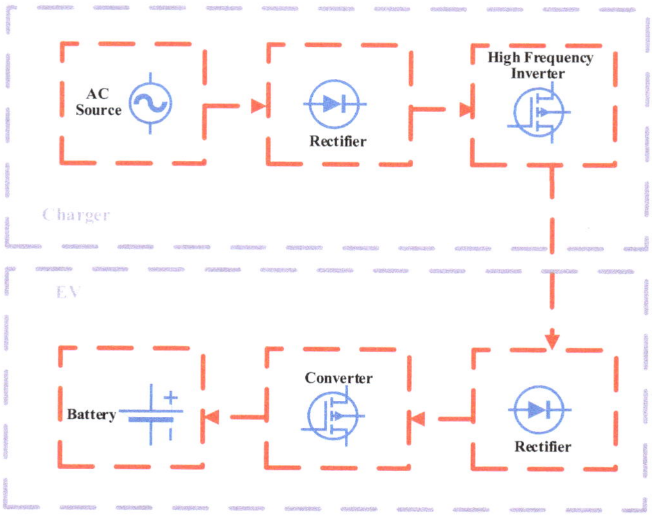

Figure 2.7 Simple configuration of inductive charging

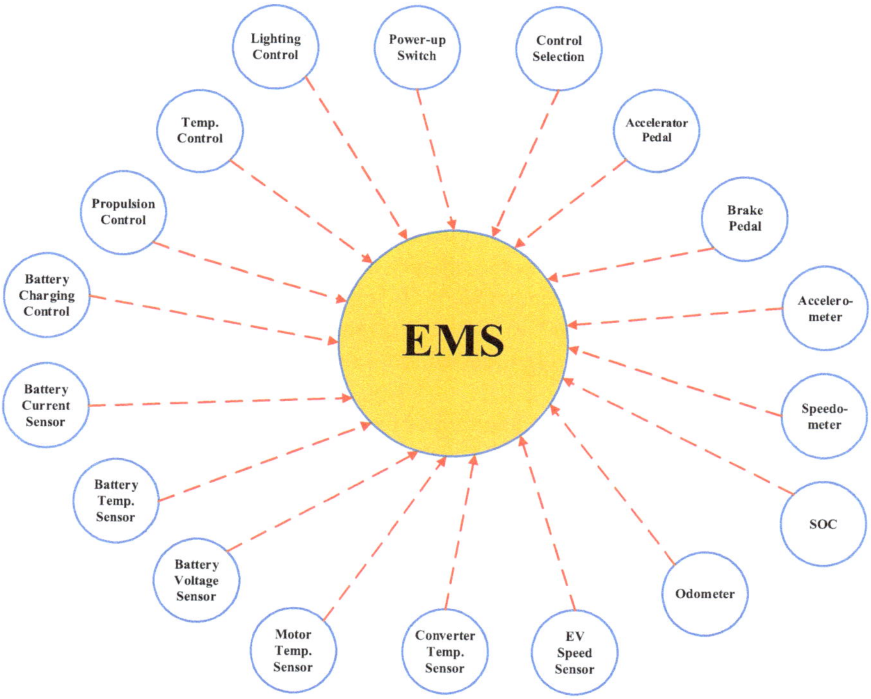

Figure 2.8 EMS configuration

current, and temperature in addition to control systems of battery charging, propulsion, and temperature.

2.3.4.1 Information exchange of EVs

Conventional charging either AC or DC follows the concept of using EVSE to supply the EV by the specified value of the charging power. However, smart charging infrastructure utilizes the power exchange concept between EVs so that the fully charged vehicle can charge the empty battery-based EV [38]. This scheme is called vehicle to vehicle (V2V) communication scheme which can be boosted to the level between EV and the main grid and named Vehicle to Grid (V2G) scheme. In this scheme, the vehicle can support the surplus power to the grid to balance the generation/demand during peak hours. Moreover, V2X is a general communication scheme that permits information and power exchange between the EV with residential and commercial aspects.

In addition to enabling EVs and chargers vendors open communication protocols, standardization makes it easier for them to sell their goods [39]. In this context, Open Charging Point Protocol (OCPP) is implemented to support this criterion. It begins with OCPP 1.5 since 2012 till OCPP 2.0 in 2018 with better improvements and more functionalities. Accordingly, open charging infrastructure enables EV drivers to charge anywhere. This can be made using a charging card, mobile app, cloud solution, or any global payment method such as credit or debit cards [40].

2.3.4.2 Management of charging/discharging infrastructure

The generic model of a communication protocol as per IEC 63110 defines four methodologies: interoperability, cyber security, grid integration, and scalability. Therefore, the e-mobility simplified role model defined in IEC 63110 is implemented in Figure 2.9 showing the communication interface between private and e-mobility networks. Consequently, communication standards shall cover the following perspectives [41]:

- Information exchange between e-mobility aspects such as metering data, energy consumed, and grid interaction.
- EVSE asset management in the form of maintenance, monitoring, control, firmware update, and configuration.
- Authentication payment of charging and discharging sessions.

The communication network shall fulfill some of the requirements as stipulated in the international standards. These requirements include supporting V2G, and V2X with the specified roles mentioned in the regulations of the distribution system operator. Information exchange between the local charging service management system (CSMS) and centralized CSMS must be standardized. It shall authorize the bidirectional energy transfer between e-mobility aspects in addition to the unconditional smart charging. However, cyber security rules shall be accomplished to guarantee end-to-end security among clients and the cloud server.

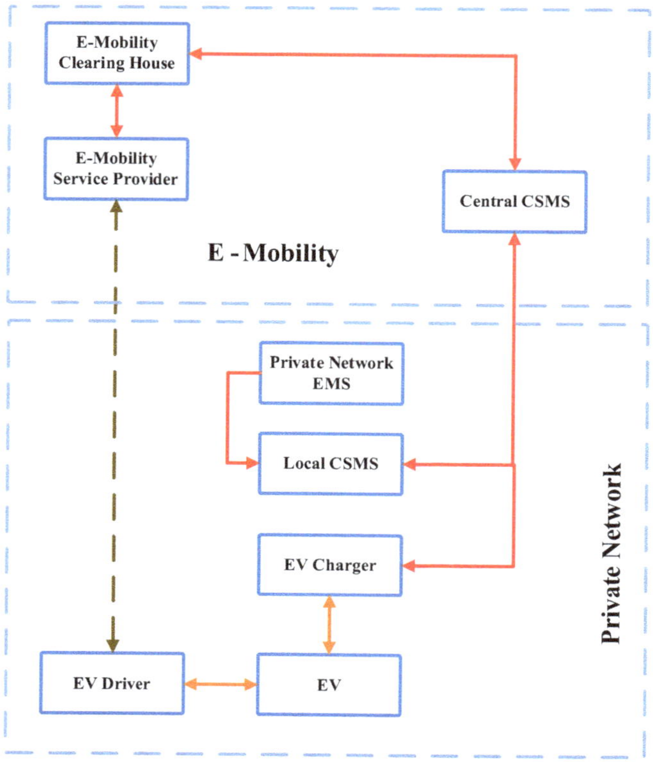

Figure 2.9 IEC 63110 representation of e-mobility network

Eventually, messaging shall support the multi-hop architecture in the scale of local and cloud platform [42].

2.4 Conclusions and future insights

In this chapter, transportation electrification platforms have been investigated with the reliance on international standards and regulations. Technical and safety requirements of batteries, electric motors, and EV auxiliaries such as connectors and plugs are also highlighted. Moreover, various charging schemes like conductive and inductive charging are interrogated clarifying their advantages and disadvantages. Standardized charging modes as stipulated in European or American standards are mentioned explaining their applications besides voltage and current ratings. However, integrating communication schemes and protocols for smart charging infrastructure is explored to enhance system connectivity and resilience. Furthermore, protection requirements of various charging schemes are surveyed to complete the portfolio of reliable and secure systems. Accordingly, this chapter proposes the guidelines and rules followed in the transportation electrification market considering local and global regulations. Eventually, further attempts shall be dedicated in the future work for investigating the limitations, challenges, and benefits of electrifying the transportation sector.

References

[1] Waleed Alabri and Dilan Jayaweera, "Optimal coordination of unbalanced power distribution systems with integrated photovoltaic systems and semi-fast electric vehicles charging stations", *IET Generation, Transmission & Distribution*, 2022, 16(12), 2399–2415, doi:10.1049/gtd2.12458.

[2] Reza Sa'adati, Meysam Jafari-Nokandi and Javad Saebi, "Allocation of RESs and PEV fast-charging station on coupled transportation and distribution networks", *Sustainable Cities and Society*, 2021, 65, 102527, doi:10.1016/j.scs.2020.102527.

[3] Zhang Qian, Zhu Yi, Wang Zhong, Hu Yue and Su Yaojia, "Siting and sizing of electric vehicle fast-charging station based on quasi-dynamic traffic flow", *IET Renewable Power Generation*, 2021, 14(19), 4204–4214, doi:10.1049/iet-rpg.2020.0439.

[4] Hossam A. Gabbar, Ahmed M. Othman, Francesco Pino and Maurizio Repetto, "Improved performance of flywheel fast charging system (FFCS) using enhanced artificial immune system (EAIS)", *IEEE Systems Journal*, 2020, 14(1), 824–831, doi:10.1109/JSYST.2019.2892002.

[5] Xiaoyu Duan, Huimiao Chen, Yiwen Song, Zechun Hu and Yonghua Song, "Planning of plug-in electric vehicle fast-charging stations considering charging queuing impacts", *IET Smart Grid*, 2020, 3(6), 786–793, doi:10.1049/iet-stg.2020.0109.

[6] Ahmed M. Othman, Hossam A. Gabbar, Francesco Pino and Maurizio Repetto, "Optimal electrical fast charging stations by enhanced descent gradient and Voronoi diagram", *Computers & Electrical Engineering*, 2020, 83, 106574, doi:10.1016/j.compeleceng.2020.106574.

[7] Mostafa M. Mahfouz and Reza Iravani, "Mitigation of the mutual dynamic interactions between a direct-current fast charging station and its host distribution grid", *IET Generation, Transmission & Distribution*, 2022, 16(8), 1541–1551, doi:10.1049/gtd2.12269.

[8] Yacine Sehimi, Khaled Almaksour, Emilia Suomalainen and Benoit Robyns, "Mitigating the impact of fast charging on distribution grids using vehicle-to-vehicle power transfer: A Paris city case study", *IET Electrical Systems in Transportation*, 2023, 13(1), e12051, doi:10.1049/els2.12051.

[9] Mohammad Amir, Zaheeruddin, Ahteshamul Haque, V. S. Bharath Kurukuru, Farhad Ilahi Bakhsh and Akbar Ahmad, "Agent based online learning approach for power flow control of electric vehicle fast charging station integrated with smart microgrid", *IET Renewable Power Generation*, 2022, 1–13, https://doi.org/10.1049/rpg2.12508.

[10] Pasquale Franzese, Dhruvi Dhairya Patel, Ahmed A S Mohamed, *et al.*, "Fast DC charging infrastructures for electric vehicles: Overview of technologies, standards, and challenges", *IEEE Transactions on Transportation Electrification*, 2023, 9(3), 3780–3800, doi:10.1109/TTE.2023.3239224.
[11] Abdelmonem Draz, Ahmed M. Othman and Attia A. El-Fergany, "State-of-the-art with numerical analysis on electric fast charging stations: Infrastructures, standards, techniques, and challenges", *Renewable Energy Focus*, 2023, 47, 100499, doi:10.1016/j.ref.2023.100499.
[12] Md Ahsanul Hoque Rafi and Jennifer Bauman, "A comprehensive review of DC fast-charging stations with energy storage: Architectures, power converters, and analysis", *IEEE Transactions on Transportation Electrification*, 2021, 7(2), 345–368, doi:10.1109/TTE.2020.3015743.
[13] Waqas ur Rehman, Rui Bo, Hossein Mehdipourpicha and Jonathan W. Kimball, "Sizing battery energy storage and PV system in an extreme fast charging station considering uncertainties and battery degradation", *Applied Energy*, 2022, 313, 118745, doi:10.1016/j.apenergy.2022.118745.
[14] S. Hemavathi and A. Shinisha, "A study on trends and developments in electric vehicle charging technologies", *Journal of Energy Storage*, 2022, 52, 105013, doi:10.1016/j.est.2022.105013.
[15] Akhtar Hussain, Van-Hai Bui and Hak-Man Kim, "Deep reinforcement learning-based operation of fast charging stations coupled with energy storage system", *Electric Power Systems Research*, 2022, 210, 108087, doi:10.1016/j.epsr.2022.108087.
[16] Mohsen Momenitabar, Zhila Dehdari Ebrahimi and Jeremy Mattson, "Queueing-based formulation model in a public transit network considering energy storage technology and demand charges", *Journal of Energy Storage*, 2023, 64, 107212, doi:10.1016/j.est.2023.107212.
[17] Chau C. Chan and Kwok T. Chau, *Modern Electric Vehicle Technology*, Oxford: Oxford University Press, 2001.
[18] James Larminie and John Lowry, *Electric Vehicle Technology Explained*, Chichester: John Wiley & Sons Ltd, 2003.
[19] Eliza Gagatsi, Thomas Estrup and Aristos Halatsis, "Exploring the potential of electrical waterborne transport in Europe: The E-ferry concept", *Transportation Research Procedia*, 2016, 14, 1571–1580, doi:10.1016/j.trpro.2016.05.122.
[20] IEC 60086-1:2021, Primary batteries – Part 1: General.
[21] IEC 60086-2:2021, Primary batteries – Part 2: Physical and electrical specifications.
[22] IEEE Std 11-2006, IEEE standard for rotating electric machinery for rail and road vehicles.
[23] ISO 17409:2020, Electrically propelled road vehicles – Conductive power transfer – Safety requirements.
[24] IEC 61851-25:2020, Electric vehicle conductive charging system – Part 25: DC EV supply equipment where protection relies on electrical separation.

[25] IEC 61851-1:2017, Electric vehicle conductive charging system – Part 1: General requirements.
[26] IEC 61851-23:2014, Electric vehicle conductive charging system – Part 23: DC electric vehicle charging station.
[27] IEEE Std 2030.1.1-2021, IEEE standard for technical specifications of a DC quick and bidirectional charger for use with electric vehicles.
[28] Mohammed Al-Saadi, Bartosz Patkowski, Maciej Zaremba, *et al.*, "Slow and fast charging solutions for Li-Ion batteries of electric heavy-duty vehicles with fleet management strategies", *Sustainability*, 2021, 13(19), 10639, doi:10.3390/su131910639.
[29] Aryan Shah, Khushi Shah, Charmi Shah and Manan Shah, "State of charge, remaining useful life and knee point estimation based on artificial intelligence and machine learning in lithium-ion EV batteries: A comprehensive review", *Renewable Energy Focus*, 2022, 42, 146–164, doi:10.1016/j.ref.2022.06.001.
[30] IEC 61851-21-1:2017, Electric vehicle conductive charging system – Part 21-1 Electric vehicle on-board charger EMC requirements for conductive connection to AC/DC supply.
[31] IEC 62196-1:2022, Plugs, socket-outlets, vehicle connectors and vehicle inlets – Conductive charging of electric vehicles – Part 1: General requirements.
[32] IEC 62196-2:2022, Plugs, socket-outlets, vehicle connectors and vehicle inlets – Conductive charging of electric vehicles – Part 2: Dimensional compatibility requirements for AC pin and contact-tube accessories.
[33] IEC 62196-3:2022, Plugs, socket-outlets, vehicle connectors and vehicle inlets – Conductive charging of electric vehicles – Part 3: Dimensional compatibility requirements for DC and AC/DC pin and contact-tube vehicle couplers.
[34] IEC TS 62196-4:2022, Plugs, socket-outlets, vehicle connectors and vehicle inlets – Conductive charging of electric vehicles – Part 4: Dimensional compatibility and interchangeability requirements for DC pin and contact-tube accessories for Class II or Class III application.
[35] IEC 62196-6:2022, Plugs, socket-outlets, vehicle connectors and vehicle inlets – Conductive charging of electric vehicles – Part 6: Dimensional compatibility requirements for DC pin and contact-tube vehicle couplers intended to be used for DC EV supply equipment where protection relies on electrical separation.
[36] IEC 61851-21-2:2018, Electric vehicle conductive charging system – Part 21-2: Electric vehicle requirements for conductive connection to an AC/DC supply – EMC requirements for off board electric vehicle charging systems.
[37] Devendra Patil, Matthew McDonough, John Miller, Babak Fahimi and Poras T. Balsara, "Wireless power transfer for vehicular applications: Overview and challenges", *IEEE Transactions on Transportation Electrification*, 2018, 4(1), 3–37, doi:10.1109/TTE.2017.2780627.

[38] IEC 61851-24:2014, Electric vehicle conductive charging system – Part 24: Digital communication between a D.C. EV charging station and an electric vehicle for control of D.C. charging.
[39] IEC 63119-1:2019, Information exchange for electric vehicle charging roaming service – Part 1: General.
[40] IEC 63119-2:2022, Information exchange for electric vehicle charging roaming service – Part 2: Use cases.
[41] IEC 63110-1:2022, Protocol for management of electric vehicles charging and discharging infrastructures – Part 1: Basic definitions, use cases and architectures.
[42] ISO 15118-2:2014, Road vehicles – Vehicle to grid communication interface – Part 2: Network and application protocol requirements.

Chapter 3

The hybrid land and sea electric transportation applications: from concept toward current practices

Mohammad Mohsen Hayati[1], Ali Aminlou[1], Hossein Biabani[1], Hassan Majidi[1], Hadi Sadeghi[2], Mehdi Abapour[1] and Foad H. Gandoman[3]

As land and sea transport electrifies, this chapter examines the novel concept of hybrid land and sea electric transportation applications. Integrating electric propulsion technologies across multiple modes of transport is becoming more important to reduce greenhouse gas emissions and transportation system environmental impact. The chapter discusses hybrid land and sea electric transportation systems' theoretical foundations, design principles, and practical implementations. The chapter also discusses hybrid transportation systems' theoretical foundations, emphasizing land-sea synergy. It examines electric propulsion system power generation, storage, and hybridization. It also examines the economic and environmental benefits of hybrid land-sea transport, providing sustainable transportation solutions for a rapidly urbanizing world. Case studies and real-world examples of hybrid land and sea electric transportation applications show their versatility in urban commuting, cargo logistics, and maritime operations. It also discusses infrastructure requirements and other implementation issues. This chapter also highlights the transformative potential of hybrid land and sea electric transportation applications for sustainable, eco-friendly transport. These systems can reduce carbon emissions and improve global transportation by integrating and optimizing electric propulsion modes.

3.1 Introduction

The concept of hybrid land and sea electric transportation represents a great step toward a sustainable and greener future [1]. The inexorable progression of

[1]Faculty of Electrical and Computer Engineering, Energy Systems Research Institute (ESRI), Smart Energy Systems Research Lab (SeS Lab), University of Tabriz, Tabriz, Iran
[2]Planning and Research Department, Fars Regional Electric Company, Shiraz, Iran
[3]Reliability & Safety Technical Center (RSTER), Zaventem, Belgium

modern transportation systems toward sustainability and environmental responsibility has spurred the development of innovative solutions that integrate electric powertrains into both land and sea vehicles. This development embodies a strategic approach to mitigate the twin challenges of climate change and the depletion of fossil fuels [2,3]. The historical infrastructure of this transformative concept can be traced back to the late nineteenth century with the advent of electric vehicles (EVs) [4]. Over time, advancements in battery technologies, power electronics, and electric propulsion systems have integrated traditional land-based and maritime transportation. Hybrid electric transportation systems, which combine conventional combustion engines with electric propulsion, represent a crucial bridge toward a greener and more sustainable transport ecosystem [5,6]. From passenger vehicles to commercial freight carriers, from leisure boats to cargo ships, the integration of electric power into these domains brings new sight of innovations [7]. Transporting consumer goods is a crucial aspect of commerce, with nearly 80% of this transport occurring via ships [8]. Unfortunately, like many transportation systems, the maritime sector primarily relies on fossil fuels, leading to the emission of CO_2 and other harmful gases [9]. This contributes to climate change, especially in coastal regions. Hence, the maritime industry must focus on technological advancements for environmentally sustainable practices. These developments aim to reduce CO_2 and other hazardous gas emissions, subsequently enhancing public health. The maritime sector plays a vital role in the global economy, particularly in Europe, where short sea operations, including ferry services for both cargo and passengers, are prevalent [10]. Europe boasts around 900 ferries, representing over 70% of global ferry journeys [11]. These operations are crucial due to the unique geography of many European regions, which necessitates intensive ferry transportation. Unfortunately, most ferries rely on fossil fuels, resulting in emissions such as CO_2, SO_2, and NO_X [12]. Alarmingly, if no action is taken, it is predicted that the growth in SO_2 and NO_X emissions will exceed the total emissions from shipping fuel consumption by 2030 [13].

The concept of seaport microgrids (SMGs), as articulated in [1], represents a novel framework for enhancing seaport management. The primary motivation behind the establishment of an SMG is to cultivate an energy district with the overarching goal of augmenting the penetration of renewable energy sources (RES) and bolstering grid storage capacity through the sale of surplus electricity to the broader market via the primary grid infrastructure [14,15]. In addition, heightened environmental concerns have prompted the International Maritime Organization (IMO) to enact specific regulations targeting emissions. Despite the significance of ferry operations to the European maritime economy, many European ferries and vessels lack the necessary specifications to address mounting environmental issues. Some ferries operating in Europe for the past two decades have not undergone updates, making them a significant source of carbon emissions. Also, there is a growing trend toward electrifying ships through hybridization, incorporating RES and various energy storage systems (ESS) within this process [16–18].

3.2 Challenges of the electrification of land and sea transport

Ferry electrification, despite its significant importance, faces multifaceted challenges that hinder its successful integration. Here are some of the key challenges for Technologies for the Electrification of Land and Sea Transport; addressing these challenges in the discussion of worldwide trends in maritime applications for the electrification of transport is essential for a comprehensive understanding of the subject and for driving the transition toward more sustainable maritime transport systems [19,20].

Diverse maritime ecosystem: The maritime sector encompasses a wide range of vessels, including cargo ships, passenger ships, fishing boats, and more. Each of these vessel types may have unique requirements and challenges when it comes to electrification. Addressing the electrification needs of this diverse ecosystem can be complex.

Infrastructure development: To support the electrification of maritime transport, significant infrastructure developments are necessary. This includes the installation of charging or refueling stations, grid upgrades, and more. Coordinating and funding these developments on a global scale is a major challenge.

Energy storage: Large ships and vessels require substantial energy storage solutions to operate efficiently. Current battery technology may not be sufficient to meet the energy demands of larger vessels. Developing cost-effective and high-capacity ESS for maritime applications is a challenge.

Range and energy efficiency: Maritime transport often involves long journeys, and electrified vessels must have the range and energy efficiency to cover these distances. Balancing the need for energy storage with vessel performance is a significant challenge. Moreover, the brief port layovers of electric ferries present a critical challenge in terms of efficient battery charging, necessitating the development of specialized electrical infrastructure. Furthermore, the adoption of lightweight materials assumes paramount importance in mitigating the weight of electric vessels. This weight reduction, achieved through the utilization of specialized materials such as carbon fiber instead of traditional steel, is indispensable for enhancing energy efficiency. Moreover, the brief port layovers of electric ferries present a critical challenge in terms of efficient battery charging, necessitating the development of specialized electrical infrastructure. Furthermore, the adoption of lightweight materials assumes paramount importance in mitigating the weight of electric vessels. This weight reduction, achieved through the utilization of specialized materials such as carbon fiber instead of traditional steel, is indispensable for enhancing energy efficiency.

Regulatory frameworks: Maritime transport is subject to various international regulations and standards. Aligning these regulations with the electrification of vessels and ensuring compliance can be complex. Additionally, the development of consistent global standards is a challenge. Also, legal challenges are noteworthy, with energy regulations such as taxation policies, and ferry construction practices not providing sufficient support for transitioning toward environmentally friendly energy sources. In several European Union nations, green energy is subject to taxation, while hydrocarbon fuels enjoy tax exemptions.

3.3 The need for hybridization

The world is grappling with significant climate concerns; hence, the imperative to identify sustainable transportation options is crucial [21]. The evolution of green vehicles within the mobility industry has emerged as a pivotal factor in mitigating carbon emissions and achieving climate-related objectives. This advancement is a response to environmental challenges and the imperative to transition to low-carbon technology within the broader context of global climate goals. Hybrid land and sea electric transportation represents one such innovative technology that holds promise. Beyond curbing environmental pollutants, this technology has the potential to serve as an effective solution for alleviating traffic density on the roadway [22,23].

The urgency of addressing climate change is highlighted by the depletion of the carbon budget for achieving a 1.5°C climate target [24]. According to the Sixth Assessment Report of the Intergovernmental Panel on Climate Change (IPCC), the carbon budget for a high probability of staying at or below the 1.5°C target was already depleted by 2022 [25]. To combat this, there is a growing recognition of the need for carbon dioxide removal (CDR) technologies, with a requirement to remove about 500 $GtCO_2$ within the twenty-first century. Making the transition to low-carbon energy sources, particularly through affordable renewable electricity sources like solar PV and hybrid electric vehicles (HEVs), has considerably contributed to decarbonizing industry and energy. This shift is expected to be completed by 2050, opening the door for widespread CDR to start in the 2030s. The boom in low-cost renewable electricity not only makes it easier to carry out natural carbon sinks like reforestation but also creates new opportunities for afforestation in arid regions through renewable energy-powered desalination and direct air carbon capture and storage (DACCS) [26–28].

Figure 3.1 provides a visual representation of the significant sources of carbon production, highlighting the pivotal role played by the transportation sector. This graphical depiction unveils a comprehensive overview of various sectors contributing to carbon emissions, including electricity, transportation, and agriculture [29].

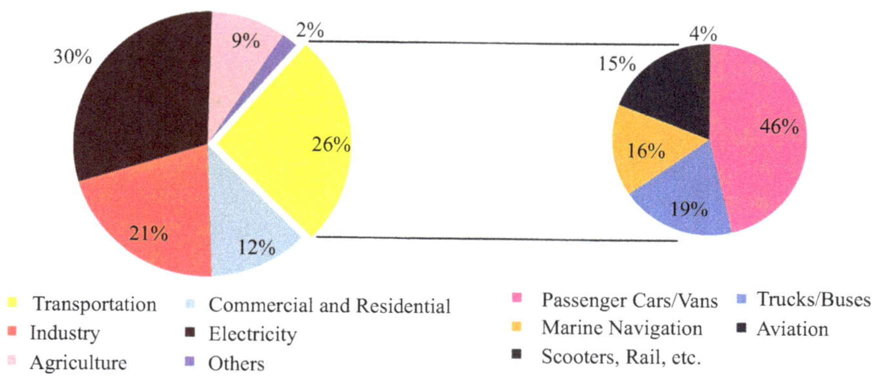

Figure 3.1 Various sector contribution to carbon emission

This figure shows the significant potential of the transportation sector in defossilization.

3.4 Land-based hybrid electric transportation

The transportation industry has recently experienced a significant transformation, driven primarily by rapid technological progress. Conventional road travel has radically changed due to the introduction of HEVs. A new frontier is emerging where the land and water domains converge. The fascinating field of land-based HEVs that are skillfully designed to traverse both terrains is examined in this subsection, providing a fascinating look at the future of amphibious transportation [30].

The combination of electric drivetrains and amphibious capabilities signifies a paradigm shift in how we think about modern transportation. These inventions offer a dynamic and diverse mode of mobility that surpasses the boundaries between land and water, surpassing the restrictions inherent in regular road travel. When one explores the world of HEVs designed for underwater navigation, it is evident that creative thinkers and cutting-edge businesses are pushing the limits of what was previously thought to be technologically impractical [31,32]. The development of vehicles like the Cybercat and Cybercat Foiler, the Trident LS-1 from Poseidon AmphibWorks, and the Multi Amphibious Vehicle (MAV) from CT&T create a framework of a future in which automobiles move between land-based and water-based thoroughfares with ease [33,34]. This part provides an entry point into a realm in which electric cars are flexible enough to transcend land-based limitations, imagining a time when commuting on land is as easy as traveling by water. Investigating these cutting-edge technologies reflects our shared dedication to environmentally benign and sustainable forms of transportation, while also attesting to human inventiveness. The possibilities presented by amphibious transportation are not limited to science fiction or specialized uses, as we stand on the cusp of this technological revolution. Instead, these advancements represent a tangible and revolutionary step forward, providing a window into a future in which electric cars coexist peacefully with open roads and urban areas.

3.4.1 Cybercat and Cybercat Foiler

The combination of amphibious capabilities and electric propulsion systems, shown by the Cybercat and Cybercat Foiler, is a novel technical effort to transform Tesla's Cybertruck into an advanced amphibious catamaran. This scholarly investigation explores the various facets of these innovative parts, clarifying their construction's subtleties and crucial function in rethinking the possibilities of amphibious mobility. The Cybertruck is a strong contender for incorporating amphibious capabilities due to its adjustable air suspension, sturdy design, and large battery capacity. The Cybercat represents a paradigm change in the adaptability of electric cars with its adjustable riding height and cleverly constructed folding inflatable pontoons that ensure ideal ground clearance on land and permit seamless water entry [35].

Anthony Diamond's idea became a reality with the Cybercat, a transportable, hydrofoiling electric watercraft that one person can assemble or disassemble. The finely engineered elements fold into discrete sections for easy storage inside the Cybertruck vault. Integrated sensors and digitally validated connections guarantee a smooth assembly process and reduce the transition time to only 5–10 minutes, highlighting this revolutionary technology's effectiveness and ease of use [36].

3.4.1.1 Evolution of the Cybercast concept

The goal of creating an accessible and eco-friendly water vehicle led Anthony Diamond to embark on a creative journey that resulted in the creation of the Cybercast. Inspired by Elon Musk's interesting tweets suggesting that the Cybertruck could be used in the water, Diamond saw the drawbacks of conventional wheel rotation and decided to come up with a novel solution. Early iterations of Diamond's proposal, which included an electric water vehicle that could hydrofoil and was movable, ran into practical and safety issues, particularly in cold climates. It soon became apparent that these obstacles could be overcome by a vehicle similar to the Cybertruck but designed for faster speeds in the water. The conversion of electric road vehicle design ideas into high-performance amphibious catamarans ensured a smooth transition from a notional personal watercraft to the realized Cybercat and Cybercat Foiler.

One notable technological feature of the Cybercat and Cybercat Foiler is the front hydrofoil's use of forward-scanning sonar sensors. This state-of-the-art technology improves submerged object identification with a detection range of over 3,000 feet. Moreover, its incorporation with car crash prevention systems enables self-driving evasive actions, highlighting the car's ability to put human safety first [37].

3.4.1.2 Environmental impact and market expansion

The Cybercat and Cybercat Foiler provide a strong argument for market expansion because they offer convenient, affordable access to recreational boating. By reducing air, water, and carbon pollution, these vehicles greatly aid in responsible environmental stewardship, which goes beyond the novelty of a new leisure experience. The Cybercat series is positioned as a key participant in furthering the worldwide drive toward sustainable energy because of its revolutionary power.

To sum up, the Cybercat and Cybercat Foiler are an example of how cutting-edge design, technological capability, and environmental awareness can come together. Their incorporation into the Tesla Cybertruck not only increases the vehicle's usefulness but also ushers in a new era of amphibious travel, representing a major advancement in the development of adaptable and sustainable EVs. Figure 3.2 illustrates the Cybercat and Cybercat Foiler, showcasing their designs as advanced amphibious electric vehicles. These models integrate Tesla's Cybertruck as their base, emphasizing electric propulsion and amphibious capabilities [36,38].

3.4.2 Poseidon AmphibWorks

With its latest creation, the Trident LS-1, Poseidon AmphibWorks Corporation, an innovative transportation technology startup co-founded by veteran developer and EV enthusiast Steve Tice and captain and amphibious vehicle specialist Andy Langesfeld, is

Figure 3.2 Images of Cybercat and Cybercat Foiler [36,38]

poised to revolutionize both land and marine travel. When it comes to amphibious vehicles, the Trident LS-1 stands out thanks to its innovative hydrofoil technology. Its operating efficiency is demonstrated by its extrapolated range of 60 miles at a speed of 25 mph on water or 22 nautical miles per hour cruising speed. By raising the Trident LS-1 above the water's surface, the hydrofoil enhancement not only increases the vehicle's range but also provides a smooth, fluid ride. This unique feature leads to a noticeable 60%–80% decrease in water drag, which is a significant improvement over traditional non-hydrofoil electric monohull boats. The strut-mounted hydrofoils skim the water well, allowing the LS-1 to reach a top speed of 35 mph without the need to deploy the hydrofoils. When the foils are fully extended, the vehicle can ride waves as high as 4 feet, which guarantees constant stability and safety in a variety of water conditions. Poseidon AmphibWorks uses a metaphor to describe the hydrofoil-enabled transition as a "magic carpet ride," which highlights how thrilling the Trident LS-1 is and how it goes beyond a simple utility vehicle to become an immersive adventure [39].

Underwater and on land, the Trident LS-1 exhibits unmatched flexibility and agility. Passengers are guaranteed an ergonomic and comfortable atmosphere with soft windows and retractable soft-tops providing weather protection. The movable covered back deck enhances the usefulness of camping by enabling customers to customize the area with elements like tables, chairs, and other fixtures. With its sliding soft covers, multipurpose tailgate, and fold-down side panels, the convertible design optimizes space efficiency and ease for a variety of uses. The Trident LS-1 can be used for "pontoon boating" on calm waters because it has a swiveling driver's seat, customizable rear decks, and room for a fourth passenger on the middle seat in back. The addition of a camping tent increases the vehicle's utility and adds comfort for longer camping trips.

Poseidon AmphibWorks believes that amphibious vehicles, like the LS-1, will play a critical role in helping people adjust to a world that is changing due to climate change. The IPCC has highlighted the apparent trend of rising sea levels, which is anticipated to increase demand for multipurpose vehicles that can navigate both land and water settings with ease. The electric propulsion of the LS-1 is in line with eco-friendly and sustainable transportation paradigms. Tice and Langesfeld envisioned the Trident as a game-changer for the recreational, ride-sharing, water taxi, cargo transport, and search-and-rescue industries. It also serves as evidence of Poseidon AmphibWorks' unwavering dedication to promoting environmentally friendly, productive, and adaptable vehicle solutions that go beyond conventional modes of transportation in areas near bodies of water [40].

In conclusion, the Trident LS-1 represents creativity and versatility and offers an insight into amphibious transportation's future that goes beyond traditional distinctions between land and sea travel. Figure 3.3 presents images that show Trident LS-1 vehicle.

3.4.3 CT&T's multi-amphibious vehicle

At the Detroit Auto Show, CT&T unveiled its MAV, a ground-breaking amalgam of amphibious features and EV technology. This revolutionary advancement calls

Figure 3.3 Trident LS-1 vehicle [40,41]

for an investigation of how well suited these cars are for urban environments, providing information on possible changes in transportation habits and the necessity of taking infrastructure into account.

The CT&T MAV upends preconceived ideas about urban transportation with its remarkable 40 mph land speed and its audacious 95 mph water speed. The notion of EVs smoothly transferring between land-based and water-based routes offers a novel remedy for traffic jams and bad weather. People who commute might think about driving over streams to avoid traffic jams or overcome obstacles caused by bad weather, which would increase the utility of electric cars in city environments. This versatility is in line with the idea of how urban transportation is evolving toward extremely efficient electric cars that function like golf buggies. The idea that commuters may alter their routes to include waterways or travel on roads free of snowdrifts emphasizes how useful these cars could be in a variety of metropolitan settings. But achieving these advantages requires a deeper comprehension of the MAV's amphibious capabilities, especially regarding the veracity of claimed water speed measurements. Despite the excitement that surrounds the CT&T MAV, a few issues and factors need to be carefully examined. Variations in stated water speed values prompt questions about the vehicle's true aquatic capabilities and call for additional clarification from the manufacturer. This misunderstanding highlights how important it is for the automobile sector to disseminate accurate and clear information, especially when it comes to cutting-edge technologies [42]. Figure 3.4 presents images that show CT&T MAV vehicle.

Figure 3.4 CT&T MAV vehicle [42]

3.5 Maritime hybrid electric transportation

One of the key developments in the maritime industry is the adoption of hybrid electric propulsion systems for ships and boats. This innovative approach combines traditional fuel-based engines with electric power, aiming to reduce emissions, enhance fuel efficiency, and improve overall environmental sustainability. Maritime hybrid electric propulsion systems typically involve the integration of conventional internal combustion engines with electric motors and ESS. This synergy allows for greater flexibility in power management, enabling vessels to operate in various modes to optimize efficiency and minimize environmental impact. The electric component can be powered by batteries, fuel cells, or a combination of both [43]. Case studies such as the Norled Ampere, MS Roald Amundsen, and projects like Yara Birkeland exemplify the potential of hybrid technology to reshape the landscape of the maritime industry.

3.5.1 Real-world applications of maritime hybrid electric transportation

3.5.1.1 Ferries

The Norled Ampere: In 2015, the world's first fully EV ferry, the Norled Ampere, started operating in Norway. It utilizes a battery pack for power, significantly reducing emissions and operating costs. The electrically propelled ferry in question is the first of its kind globally, emitting no pollutants and generating minimal noise. Launched in October 2014, it officially began its commercial services in May 2015. This cutting-edge vessel operates along a 5.7-kilometer route across the Sognefjord, linking the communities of Lavik and Oppedal in Norway [44].

Washington state ferries (WSF): WSF is actively exploring hybrid electric solutions as part of an ambitious initiative to transition its fleet of vessels to emission-free operations. As the largest ferry system in the United States, WSF recognizes the significance of incorporating hybrid propulsion technology to mitigate the environmental impact associated with its operations. Notably, WSF stands as the largest contributor to greenhouse gas emissions among all state agencies in Washington, consuming 19 million gallons of diesel fuel annually to facilitate the transportation of 24 million passengers. The decision to adopt hybrid electric ferries aligns with directives from the governor and legislature, as well as the strategic goals outlined in the agency's long-range planning initiatives. Securing approximately $1.33 billion in funding, the total estimated cost for this transformative project is $3.98 billion [45].

3.5.1.2 Cargo ships

MS Roald Amundsen: The ship represents the inaugural addition of two hybrid vessels to the fleet of Hurtigruten Expeditions, constructed at Kleven Yards in Norway in 2019. The adoption of electrical propulsion contributes to a substantial reduction in emissions. Functioning as a hybrid electric-powered expedition cruise ship, it incorporates battery packs to complement its low-emission engines, resulting in a noteworthy decrease of over 20% in CO_2 emissions when compared to other cruise ships of equivalent size [46].

Yara Birkeland: Yara International, in collaboration with Kongsberg, is developing the world's first fully electric and autonomous container ship that boasts a zero-emission profile. The vessel aims to revolutionize the shipping industry by eliminating emissions and reducing operational costs through electric propulsion. Projections indicate that this vessel will lead to a notable reduction of 40,000 annual journeys in diesel-powered truck haulage for Yara [47].

3.5.1.3 Yachts

Silent-Yachts: Silent-Yachts specializes in solar-powered catamarans and yachts. These luxury vessels feature a combination of solar panels, batteries, and conventional power sources, allowing for extended cruising ranges with minimal environmental impact. The solar-electric propulsion system employed in these yachts achieves unprecedented levels of reliability and efficiency [48]. Figure 3.5 illustrates a sample of a Silent-Yachts boat and solar panels connected to it.

Oceanco's Tuhura: While not a hybrid in the traditional sense, Oceanco's Tuhura superyacht concept incorporates advanced propulsion technology to enhance efficiency and reduce environmental impact, showcasing the integration of innovative technologies in the luxury yacht sector. The yacht is equipped with a Hybrid Contra Rotating Propeller System for its propulsion, featuring technology provided by ABB [49].

Oceanco's vessel has been developed through a collaborative effort involving Lobanov Design Studio, naval architect BMT Nigel Gee, and interior designer Achille Salvagni. While Tuhura stands out in a unique category, the CRP Azipod propulsion system has a commercial lineage. Initially chosen in 2004 for the 32-knot Shin Nihonkai Ferry Co. RoPax Ferries Akashia and Hamanasu, delivered by Mitsubishi Heavy Industries, these vessels were the largest and fastest ferries in Japan at that time. Notably, they exhibited enhanced maneuverability compared to their predecessors, accompanied by a 20% reduction in fuel consumption [50]. Figure 3.6 illustrates a sample of a Oceanco's Tuhura ship. Figure 3.7 shows the general division of maritime hybrid electric transportation case studies.

Figure 3.5 A sample of a Silent-Yachts

Figure 3.6 A sample of a Oceanco's Tuhura

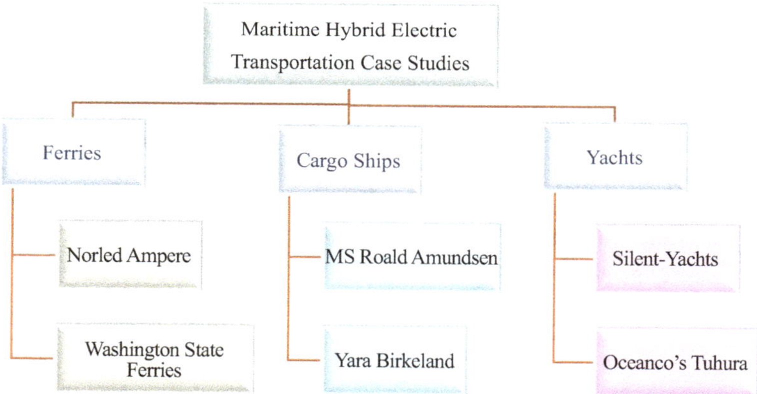

Figure 3.7 An overview of maritime hybrid electric transportation case studies

3.6 Integrating land and sea hybrid systems

To make the dream of seamless inter-modal electric transportation networks that connect ports, highways, and water pipelines come true, it will be important to make sure that navigation, vehicle-to-vehicle communication, and autonomous controls all work together well. New battery, motor, and digital technologies are likely to be very important in the creation of hybrid land-sea electric platforms. These platforms could help create more connected and eco-friendly transportation systems along coastlines and between communities on land and those on islands around the world. In this situation, Yusuf's [51] study looks into how plug-in EV affect a distribution level microgrid. This study shows how complicated it is to add EVs to transportation systems. Also, Schuss *et al.* [52] show how important it is to know the right conditions for adding photovoltaic cells to passenger vehicles. This shows how important it is to use advanced PV simulation models for EV with photovoltaic installations in the context of hybrid land-sea electric transportation applications. These sources show how important research and new technologies are for shaping the future of hybrid electric transportation systems, especially when it comes to combining EV with other types of transportation.

Overall, combining land and sea hybrid electric transportation systems has a lot of potential to change the way people and goods get around in coastal areas and on different types of terrain. This could lead to more connected, environmentally friendly transportation networks that make it easier to move people and goods quickly. As battery, motor, and digital technologies keep getting better, it will soon be possible to have inter-modal electric transportation networks that run on both land and sea. This will make transportation around the world more sustainable and connected. Authors of [1] provided valuable insights into two promising technologies that are enabling more sustainable maritime transportation: SMG and all-electric ships (AES). The paper discussed how electrifying both seaports and ships can enhance operational flexibility and energy efficiency while reducing environmental impacts, as electrification is transforming connectivity between them from solely logistical to also spanning the electric domain via technologies like cold-ironing. This changes maritime transportation management into a complex multi-microgrid coordination problem. To address this, the authors proposed a hierarchical control framework for future seaport management, demonstrating its validity using a hardware-in-the-loop experiment of active synchronization between a shipboard DC microgrid and seaport AC microgrid to help mitigate impulse power issues from plug-in events.

While hybrid land and sea electric transportation presents many opportunities to reduce emissions and dependence on fossil fuels compared to traditional gas-powered vehicles, there are also several technological and infrastructure challenges that must be addressed. By combining electric battery and motor technology with efficient diesel or gas engines, hybrid vehicles can utilize electricity for basic transportation needs yet still power accessory functions and extend the driving range over long distances to more practical electric-only designs. This flexibility allows hybrid designs to work for applications ranging from passenger cars and trucks to larger ships and ferries traveling between ports. However, developing highly efficient yet durable hybrid powertrains that optimize both fuel and electrical use over the full lifecycle of vehicles requires extensive research and testing. Building out networks of charging stations along highway corridors and near ports is another massive undertaking that demands coordination between government agencies and private industry. Standardizing electrical specifications, high-capacity batteries, and fast-charging technology present compatibility and interoperability issues to resolve. Making hybridization economical overall compared to conventional vehicles depends upon achieving scale in production to drive down costs through competition and innovation. Overcoming these hybrid transportation challenges can significantly reduce harmful emissions in the coming decades and help transition toward eventual electrification of the entire transportation sector while maintaining necessary ranges and duties in the interim through the optimized use of multiple energy sources.

Unified demonstrations and pilot programs allow for streamlined research, optimization of solutions, leveraging of combined resources, and more rapid scaling. Some insightful project collaborations that exemplify this approach include the following: the possibility of using hybrid electric propulsion systems for inland barges in Poland is studied in [53]. It provided useful context on the history and current state of inland water transportation in the country. The authors outlined Poland's plans to upgrade its waterways according to AGN (Agreement on Main Inland Waterways of International Importance) Convention standards and highlights obstacles like the poor quality of many routes. Additionally, the paper presented various historical electric propulsion designs that were implemented on inland vessels in the early twentieth century. It then examined the energy efficiency and emissions benefits of different transportation modes like inland shipping versus road or rail transport. Importantly, the author concluded that hybrid and electric solutions can significantly reduce fuel use and emissions from inland vessels.

3.7 Sustainability in the expectation of emission mitigation

As can be observed from the current motivations of investment firms, it seems that burgeoning the hybrid land and sea electric transportation concept is related to the evolution of modern world, amalgamated with a diverse spectrum of demands, pleasures, and tendencies, rather than environmental issues and optimization of transportation systems. The reasons behind such inference can be enumerated as follows:

- Realization of the hybrid land and sea electric transportation systems (HLSETS) aligned with business objectives on large scales seems not to be economical at least for coming decades, whereas maritime transportation infrastructures as well as land transportation ones under optimal operating conditions being improved are available in almost all over the world. Therefore, employing HLSETS only in some specific application fields, such as military area, seems to be affordable.
- As the application area of HLSETS is limited and specific, environmental considerations cannot be the first and powerful arguments for their development and diffusion.
- Regarding the growing electrification trend of maritime transportation systems, especially in small-scale vessels, in line with promotion of renewable-based land electric transportation systems, the advent of HLSETS won't have remarkable impacts on environmental effects separately expected from the electric version of land and water transportation modes.

Accordingly, evaluating the impacts of HLSETS from the viewpoint of GHG emission will be worthy only in the application area in which HLSETS are widely used. Of awaiting application areas, military industries seem more enthusiastic about developing HLSETS. Of course, as mentioned, promotion of HLSETS, even among industries enthusiastic about it, only in response to the emission restrictions imposed on the transportation systems does not seem reasonable, because:

- Apart from navy transportation systems, military land vehicles have limited mobility during limit periods compared with passenger cars; under such circumstances, assessing their emission amounts with respect to projected war/maneuver conditions seems to be worthless.
- Military transportation systems, during more than several decades' record of progress, have absorbed heavy budgets. In this context, electric propulsion systems will be probably affordable for next new generations, and replacing current systems that have adopted required standard levels by passing quality control processes will not be on the agenda.
- Of outstanding features of military land vehicles, the reliability plays a vital role. Regarding the restrictions derived from supplying power or from required infrastructures, such as charge stations, under different conditions, electrification of the vehicles will perhaps need to more strategic and stronger reasons than environmental considerations.

It is noteworthy that above-reviewed reasons have not resulted in stopping researchers' development in the field of employing electric propulsion systems as well as diffusing HLSETS. Obtained from different battle scenes, scrutinizing the past experiences reveals that the transition of military vehicles in next generations from current fossil fuel-based engines to new electric propulsion systems along with penetration of HLSETS can reinforce military power of countries from the below view points.

- The diversity of available technologies for penetration and identification goals in military operations.
- Increase in the onrush speed affected by the possibility of simultaneous movements over land and water.
- Declining the risk of being identified via thermal vision devices and weapons systems that rely on heat signatures, while electric propulsion systems have a quieter, softer, and cooler activity.
- Replacing the fossil fuels with RES and the capability of supplying power locally. In this case, experience obtained from the Afghanistan war shows that provision of every fuel gallon in that war costs seven fuel gallons for the US Army [54].

Above reasons, together with the enhancement of energy efficiency, lower operation and maintenance cost, and environmental compatibility, indeed, have provided the required motivation for exploring and developing hybrid land and water transportation systems equipped with electric propulsion systems in military industries more than other application areas [55]. It should be noted that, from different points of view, such as sensitive operation conditions, lack of sufficient infrastructures, fire hazard, or the environmental damage potential originated from released/sunk batteries in the nature/seas, battery-based EV are still faced by some weakness points. And, as negative aspects of the context, they should be involved in assessments.

3.8 Technological challenges and future trends

Proportional to expectations from HLSETS, these systems are faced by the challenges of both transportation modes, i.e., land EVs and new generation of maritime small-scale transportation systems. From a general perspective, the HLSETS challenges can be investigated in three fields as follows.

- ***The technology field***: apart from advancements in hybrid land-sea and air-sea electric transportation systems, the current monopoly on electric propulsion technologies held by some developed countries has limited accessibility for developing nations. This restriction has hindered the maturity and even the emergence of this new paradigm of transportation systems in many other regions. In this regard, as the International Energy Agency projections show, the penetration rate of electric cars will reach to 20% by the end of 2023 [56], while the beginning of road transportation systems' electrification goes back to more than two decades ago. Lack of production and/or operation technologies is one of the reasons behind this slow penetration rate. This deficit is followed by the fear of staying without charge or without expert repairman. Subsequently, some of remarkable barriers to HLSETS should be sought in the technology field.

Being hesitant about completely realizing the prophecy of EVs in developing countries can be treated as another anxiety over their penetration levels. As we know, the concept of transportation systems' electrification has been built upon the sustainable development concept. In this context, the lack of sufficient RES in electricity generation mixes to meet demands derived from EVs can negatively affect the improvements made to overcome environmental damages obtained from sustainable development of other energy-related sectors. Hence, in such societies, challenges faced by RES employment should also be added to the set of obstacles to HLSETS. But the most important challenge of the technology field is linked to batteries, known as the main factor realizing the transition of transportation systems toward decarbonization. The reported barriers to development and progress pathways of batteries in different countries can be evaluated from two technical and non-technical (economic) viewpoints. The limit access to required initial material, such as natural cobalt sources [57], achieving higher capacity to weight (and/or volume) ratio as well as more environmental compatibility, and facilitating operating conditions by enhancing operating flexibility, are the main linchpins of current researchers' efforts in battery industries.

Beyond the enumerated challenges, i.e., the monopoly of technology, low penetration rate of RES, and improvements of batteries, the lack of infrastructures required to employ HLSETS can be considered in the next level of barriers.

The infrastructures can range from regulatory frameworks, such as maritime traffic regulations, to submarine cables, charge/swapping stations [58].

- ***The economic field***: Undoubtedly, elimination of barriers linked to the technology field increases upfront cost of HLSETS. Hence, similar to the fields of land EVs and ESS [59], it is expected that hybrid land-sea EVs are faced with the challenge of high upfront cost at least for coming years. Obviously, expanding application area of this generation of transportation systems will be contingent upon being competitive with other transportation options such as electric vertical take-off and landing (eVTOL) vehicles [60] or flying boats/airships [61] in parallel with ongoing development of electric maritime transportations systems as well as land transportation ones.
- ***The social field***: Of the first advancement conditions of each technology, scrutinizing the target market and social acceptance are crucial. As mentioned before, the backbone of defined goals for electrification of transportations systems is realization of the sustainable development. But, as can be observed from recent studies carried out in some countries such as Canada and Brazil [62,63], the level of attempts to enhance the societies' aware of the sustainability importance and to provide the capacity of acceptance is lagging behind the level of attempts to progress technologies and to focus on manufacturing hybrid land and sea EVs or eVTOL ones. In this context, people resistance to new technologies before passing their maturity phases can be treated as other reasons behind low acceptance level of EVs, in particular hybrid land and sea EVs, in various application fields. And investigations show that many of potential consumers have taken a wait-and-watch approach to switch to EVs [64]. Accordingly, refocusing on incentive mechanisms designed for RES promotion, emission mitigation measures imposed to different sectors, and on tax policies along with more attempts to increase the public awareness of the necessity of sustainable development can accelerate the future development process of HLSETS.

As a field needing highest investment cost, the technology field has always received more attention corresponding to its important and expenses. Accordingly, to cope with above-mentioned challenges, progresses in this field, ranging from battery industries to control strategies and exploitation of RES, have been significant in recent years. Among these progresses:

- ***Advancements in batteries***: Recent impressive advancements are linked to battery industries, making EVs development possible. Of these progresses, the advent of solid-state batteries has made researchers full of hope for overcoming the problems associated with utilization of current generation of batteries, i.e., lithium-ion batteries. Higher capacity to volume ratio, elimination of the fire hazard, and so the need to have components for safety, shorter charging time along with longer lifetime have been reported as the advantages of this new type of batteries compared with the current generation. But what has still kept the solid-state batteries in the research and development phase is related to ongoing efforts to address some technical issues such as the electric double layer effect [65] or the need for fast-conducting solid electrolytes to provide sufficient transport in composite cathodes [66]. Of course, in parallel with the advent of new technologies in the batteries field, EV industries have also witnessed significant progresses in the lithium-ion batteries area [67]. Improvements in storing capability, declining the memory effect [68], and increases in lifetime as well as environmental compatibility are the recent achievements in that field.

- ***Enhanced propulsion systems and control strategies***: Growing tendencies to design more optimal marine transportations systems with higher reliability in line with reaching smart ships, nowadays have led to a variety of propulsion systems based on different control strategies and power supply architectures. To better appreciate the extent of this variety, in-progress fields of this context are illustrated in Figure 3.8 The diversity of propulsion systems as well as control strategies, shown in this figure, has made operating conditions more flexible, resulting in the diversity of application areas and possible combinations. The concept of HLSETS is an example of this context.
- ***Available RES options and their growing penetration***: Among all renewable RES, solar and wind units have always received more attentions than other sources as the result of their high level of availability. Operation planning of these units in different geographical locations also has been always faced by various challenges such as provision of required infrastructures and efficiency improvement. Obviously, in some specific rainy and coastal regions, other forms of RES, such as tidal current energy, can provide greater exploitation even for lower investment cost. Hence, in parallel with the development of electric marine transportation systems in various scales and applications, investigations on development of tidal current energy converters have been accelerated. In this context, to pass commercial maturity process, construction of large-scale turbines is under study in different countries, especially along the countries highly populated coasts, such as China [69].
- ***Marine microgrids development***: In today's energy-dependent world, maximization of energy accessibility is introduced as one of the main conditions for the social welfare (SW) realization in societies. In such framework, developing infrastructures of energy systems and relevant carriers seems inevitable. And infrastructure expansion projects are in big-budget program class. Under such circumstances, advancements in distributed generation units, in particular in RES-based ones, along with the achievements of the power electronic converter field, smart control systems, and above all, of the ESS field, can end in economic approaches to making energy accessible and realizing the SW, if islanded microgrids are implemented. In this new paradigm of power systems, the need to develop hybrid land and sea EVs is more felt. In addition, employing HLSETS can be a step forward in normalization of lives in islands on the principles of the sustainable development concept.

The pursued framework, including progresses of battery industries, propulsion and advanced control systems, marine RES and microgrids, will lead to more attract researchers' attention to HLSETS in a near future. With regard to the technology

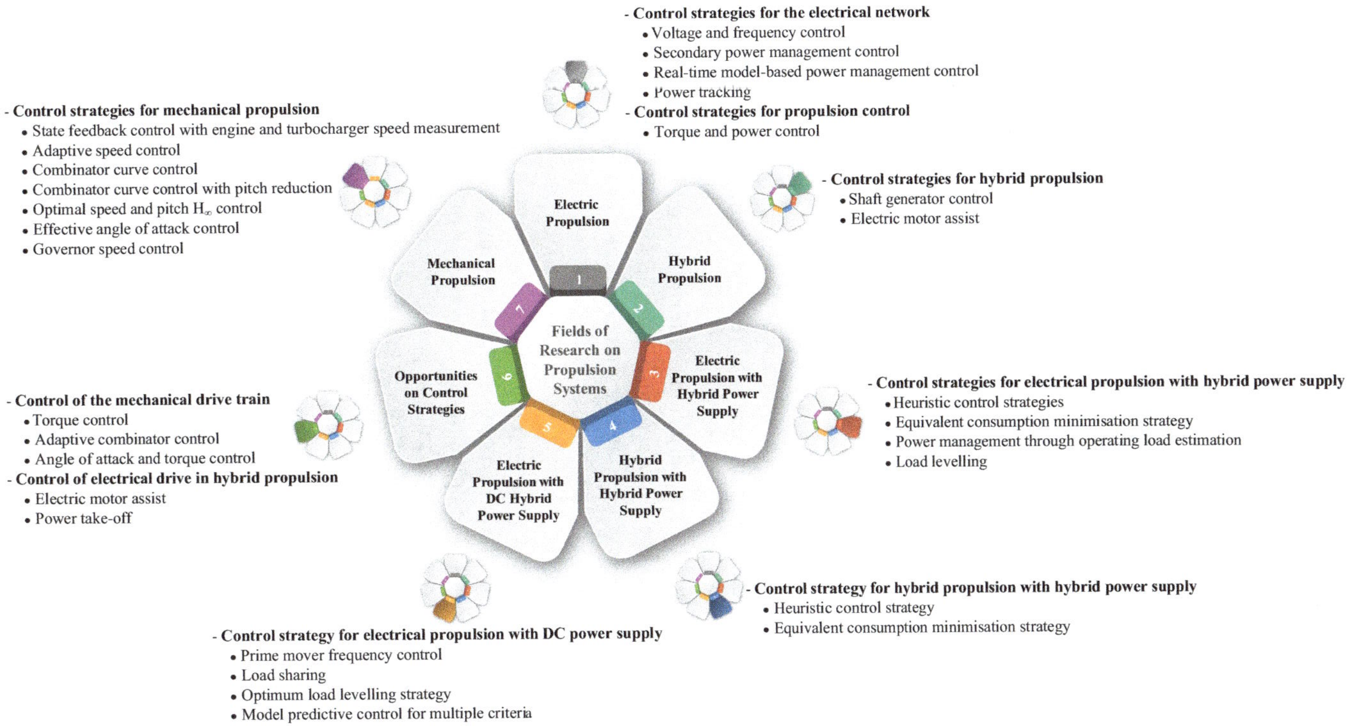

Figure 3.8 In-progress fields of propulsion systems

progress speed, in particular in hybrid modes of transportations systems, it is expected that the commercialization process of HLSETS is accomplished for coming years at least in special application fields. Of these fields, the followings are outstanding.

- ***Military operations***: As mentioned, military industries will be probably the first institutions welcoming HLSETS. With respect to the improved reliability, increased penetration and camouflage power, and being better adaptable to unpredictable operating conditions, it is projected that the utilization of HLSETS in the military field is rapidly expanded. As a matter of course, this expansion can be important from the viewpoint of sustainable development, too. Findings of the recent study carried out on electrification of the US Army military vehicles reveal that the electrification can provide the same environmental benefit of taking around 700,000 passenger vehicles off the road today [70].
- ***Rescue and relief activities***: Under some urgent and specific circumstances, the time can be the sole criterion determining the type of transportation system. Obviously, in such cases, the aviation transportation systems will be the first choice. But, in the lack of air vehicles, such as helicopters, or lack of required operating conditions, hybrid land and sea EVs can be an appropriate alternative.
- ***Research stations in the Arctic***: Regarding the existing limitations of infrastructures required to transmit energy to research statins in the arctic, and the need to move over both water and ice surfaces, combination of HLSETS and RES-based units can be a superior option for replacing.
- ***Energy trading between Archipelago islands***: In line with the development of marine microgrids in the islanded mode, developing and employing the HLSETS provides the possibility of not only movements between islands but also energy trading between them. And the latter can improve operation economy and also power supply reliability of islands [71]. This context is worthy of future study and investigation, in particular in case of islands that have harsh weather and/or are far away from energy distribution networks. For such archipelago or islands chain, the possibility of laying submarine cables doesn't exist because of the lack of engineering conditions.
- ***Wildlife watching tourism***: In many of the world countries, such as the Australia and the African countries, the tourism industry plays a significate role in growing the total GDP generated by all industries. Among tourism travel types, the wildlife watching tourism appeals to a much wider range of people than the more specialist forms of eco-tourism. This type of tourism in fact is of the most expensive travel tourism types, and hence, the market for specialist wildlife watching holidays continues to grow. In this context, through the process of watching and recording attractive scenes, turning off the engines of vehicles is a natural behavior required, as the noise of vehicles' engines may disrupt the process. Under such circumstances, replacing conventional safari vehicles with hybrid land and sea electric ones can not only facilitate commitment to the tour requirements but also provide more exciting moments via crossing over lakes or rivers and water between islands.

3.9 Conclusions

The electrification of land and sea transport is instrumental in addressing pressing environmental concerns and advancing the future of transportation. The chapter begins by establishing the theoretical foundations and design principles of hybrid transportation systems, emphasizing the synergy between land and sea travel. It highlights the potential of these systems to revolutionize global transportation by reducing greenhouse gas emissions and minimizing environmental harm. This study delves into the technical aspects of electric propulsion systems, discussing their efficiency, power generation, and storage, and how these factors contribute to hybridization. The economic and environmental benefits of adopting hybrid land and sea transport are thoroughly examined, offering insights into sustainable transportation solutions for an increasingly urbanized world. The inclusion of case studies and real-world examples demonstrates the versatility of hybrid electric transportation applications in various scenarios, from urban commuting to cargo logistics and maritime operations.

Despite the transformative potential, the chapter acknowledges the challenges faced in implementing hybrid systems, particularly in the maritime sector. These challenges include the diverse maritime ecosystem, the need for significant infrastructure development, energy storage solutions for large vessels, and the balancing of range and energy efficiency for maritime transport. Additionally, regulatory frameworks and legal challenges pose obstacles to the widespread adoption of hybridization in the maritime industry. The chapter emphasizes the urgent need for sustainable transportation options, given the serious climate concerns and the depletion of the carbon budget for achieving climate targets. It discusses the role of HEVs in lowering carbon emissions and meeting global climate goals, particularly in the context of the transportation sector's potential for decarbonization.

The integration of hybrid systems is deemed essential, and the chapter advocates for the synergy between land-based and maritime mobility. It concludes by addressing the environmental impact and sustainability of hybrid land and sea electric transportation, recognizing the importance of emission mitigation. The technological challenges and future trends in this field are also outlined, pointing toward ongoing advancements and the continuous evolution of hybrid electric transportation systems as a promising pathway toward a cleaner and more efficient future for global transportation.

References

[1] S. Fang, Y. Wang, B. Gou, and Y. Xu, "Toward future green maritime transportation: An overview of seaport microgrids and all-electric ships," *IEEE Transactions on Vehicular Technology*, vol. 69, no. 1, pp. 207–219, 2019.

[2] O. B. Inal, J.-F. Charpentier, and C. Deniz, "Hybrid power and propulsion systems for ships: Current status and future challenges," *Renewable and Sustainable Energy Reviews*, vol. 156, p. 111965, 2022.

[3] S. Khalili, E. Rantanen, D. Bogdanov, and C. Breyer, "Global transportation demand development with impacts on the energy demand and greenhouse gas emissions in a climate-constrained world," *Energies*, vol. 12, no. 20, p. 3870, 2019.
[4] M. Ehsani, Y. Gao, S. Longo, and K. Ebrahimi, *Modern Electric, Hybrid Electric, and Fuel Cell Vehicles*, 2nd edn. CRC Press, Boca Raton, FL, 2018.
[5] W. P. Symington, A. Belle, H. D. Nguyen, and J. R. Binns, "Emerging technologies in marine electric propulsion," *Proceedings of the Institution of Mechanical Engineers, Part M: Journal of Engineering for the Maritime Environment*, vol. 230, no. 1, pp. 187–198, 2016.
[6] D. Sperling, *Future Drive: Electric Vehicles and Sustainable Transportation*. Island Press, Washington, DC, 2013.
[7] T. T. Taefi, J. Kreutzfeldt, T. Held, and A. Fink, "Supporting the adoption of electric vehicles in urban road freight transport—A multi-criteria analysis of policy measures in Germany," *Transportation Research Part A: Policy and Practice*, vol. 91, pp. 61–79, 2016.
[8] J. J. Corbett and H. W. Koehler, "Updated emissions from ocean shipping," *Journal of Geophysical Research: Atmospheres*, vol. 108, no. D20, pp. 4650–4666, 2003.
[9] A. Al-Enazi, E. C. Okonkwo, Y. Bicer, and T. Al-Ansari, "A review of cleaner alternative fuels for maritime transportation," *Energy Reports*, vol. 7, pp. 1962–1985, 2021.
[10] H. Xing, C. Stuart, S. Spence, and H. Chen, "Alternative fuel options for low carbon maritime transportation: Pathways to 2050," *Journal of Cleaner Production*, vol. 297, p. 126651, 2021.
[11] A. Cuculić, L. Draščić, I. Panić, and J. Ćelić, "Classification of electrical power disturbances on hybrid-electric ferries using wavelet transform and neural network," *Journal of Marine Science and Engineering*, vol. 10, no. 9, p. 1190, 2022.
[12] D. Stenersen and O. Thonstad, "GHG and NOx emissions from gas fuelled engines: Mapping, verification, reduction technologies," Report No. OC2017 F-108, SINTEF Ocean AS, pp. 1–52, 2017.
[13] A. Vangelis, "*Towards greener shipping: Regulatory measures on ship energy efficiency and emissions reduction*." University of Piraeus (Greece), Athens, 2021.
[14] N. N. A. Bakar, J. M. Guerrero, J. C. Vasquez, *et al.*, "A review of the conceptualization and operational management of seaport microgrids on the shore and seaside," *Energies*, vol. 14, no. 23, p. 7941, 2021.
[15] M. U. Mutarraf, Y. Terriche, M. Nasir, *et al.*, "A decentralized control scheme for adaptive power-sharing in ships based seaport microgrid," in *IECON 2020 The 46th Annual Conference of the IEEE Industrial Electronics Society*, 2020, pp. 3126–3131: IEEE, Piscataway, NJ.
[16] P. Serra and G. Fancello, "Towards the IMO's GHG goals: A critical overview of the perspectives and challenges of the main options for decarbonizing international shipping," *Sustainability*, vol. 12, no. 8, p. 3220, 2020.

[17] T. C. Van, J. Ramirez, T. Rainey, Z. Ristovski, and R. J. Brown, "Global impacts of recent IMO regulations on marine fuel oil refining processes and ship emissions," *Transportation Research Part D: Transport and Environment*, vol. 70, pp. 123–134, 2019.

[18] Ø. Endresen, E. Sørgård, J. K. Sundet, *et al.*, "Emission from international sea transportation and environmental impact," *Journal of Geophysical Research: Atmospheres*, vol. 108, no. D17, p. 4560, 2003.

[19] O. Oloruntobi, K. Mokhtar, A. Gohari, S. Asif, and L. F. Chuah, "Sustainable transition towards greener and cleaner seaborne shipping industry: Challenges and opportunities," *Cleaner Engineering and Technology*, vol. 13, p. 100628, 2023.

[20] S. Anwar, M. Y. I. Zia, M. Rashid, G. Z. D. Rubens, and P. Enevoldsen, "Towards ferry electrification in the maritime sector," *Energies*, vol. 13, no. 24, p. 6506, 2020.

[21] S. Sultana, D. Salon, and M. Kuby, "Transportation sustainability in the urban context: A comprehensive review," *Urban Geography*, vol. 40, no. 3, pp. 279–308, 2019.

[22] H. G. Grimes-Casey, G. A. Keoleian, and B. Willcox, "Carbon emission targets for driving sustainable mobility with US light-duty vehicles," *Environmental Science & Technology*, vol. 43, no. 3, pp. 585–590, 2009.

[23] N. Rigogiannis, I. Bogatsis, C. Pechlivanis, A. Kyritsis, and N. Papanikolaou, "Moving towards greener road transportation: A review," *Clean Technologies*, vol. 5, no. 2, pp. 766–790, 2023.

[24] Y. D. Priore, G. Habert, and T. Jusselme, "Global carbon budgets for the built environment: How far are we to achieve a 1.5°C limit in global warming? A Swiss example," in R. Azari and A. Moncaster (eds.), *The Routledge Handbook of Embodied Carbon in the Built Environment*. Routledge, Oxfordshire, 2023, pp. 103–129.

[25] H. Lee, K. Calvin, D. Dasgupta, *et al.*, "IPCC, 2023: Climate Change 2023: Synthesis Report, Summary for Policymakers. Contribution of Working Groups I, II and III to the Sixth Assessment Report of the Intergovernmental Panel on Climate Change [Core Writing Team, H. Lee and J. Romero (eds.)]. IPCC, Geneva, Switzerland," 2023.

[26] C. Breyer, D. Keiner, B. W. Abbott, *et al.*, "Proposing a 1.0°C climate target for a safer future," *PLOS Climate*, vol. 2, no. 6, p. e0000234, 2023.

[27] H. Liu, J. D. Ampah, S. Afrane, H. Adun, C. Jin, and M. Yao, "Deployment of hydrogen in hard-to-abate transport sectors under limited carbon dioxide removal (CDR): Implications on global energy-land-water system," *Renewable and Sustainable Energy Reviews*, vol. 184, p. 113578, 2023.

[28] V. Harvey, M. Workman, and R. Heap, "Developing carbon dioxide removal policy and anticipatory perspectives in the United Kingdom and United States," *Energy Research & Social Science*, vol. 102, p. 103185, 2023.

[29] S. Qazi, P. Venugopal, G. Rietveld, *et al.*, "Powering maritime: Challenges and prospects in ship electrification," *IEEE Electrification Magazine*, vol. 11, no. 2, pp. 74–87, 2023.

[30] A. Ziółkowski, P. Fuć, A. Jagielski, M. Bednarek, and S. Konieczka, "Comparison of the energy consumption and exhaust emissions between hybrid and conventional vehicles, as well as electric vehicles fitted with a range extender," *Energies*, vol. 16, no. 12, p. 4669, 2023.

[31] X. Xiang, Z. Niu, L. Lapierre, and M. Zuo, "Hybrid underwater robotic vehicles: The state-of-the-art and future trends," *Hkie Transactions*, vol. 22, no. 2, pp. 103–116, 2015.

[32] M. Zhang, H. Gu, and J. Guo, "Energy management strategies for hybrid ships and unmanned underwater vehicles," *Journal of Physics: Conference Series*, vol. 2395, no. 1, p. 012051, 2022, IOP Publishing, Bristol.

[33] *Amphibious electric vehicles by Poseidon AmphibWorks*. Available: https://www.designboom.com/technology/amphibious-electric-three-wheeler-trident-ls-1-poseidon-amphibworks-10-17-2022/ (Accessed November 30, 2023).

[34] *Poseidon AmphibWorks Corporation*. Available: https://poseidonamphibworks.com/ (Accessed November 30, 2023).

[35] *Transforming Tesla's cybertruck into an amphibious, high-performance watercraft*. Available: https://www.designboom.com/technology/transforming-teslas-cybertruck-02-11-2022/ (Accessed November 10, 2023).

[36] *Introducing the Cybercat and Cybercat Foiler*. Available: https://www.cybercatamaran.com/ (Accessed November 10, 2023).

[37] *Tesla Cybercat Foiler*. Available: https://www.afloat.com.au/2022/02/17/tesla-cybercat-foiler/ (Accessed November 10, 2023).

[38] *Cybercat and Cybercat Foiler*. Available: https://wordlesstech.com/cybercat-and-cybercat-foiler/ (Accessed November 09, 2023).

[39] *Poseidon AmphibWorks is combining cars, boats and hydrofoils for the next new mode of transportation*. Available: https://www.thelog.com/news-departments/poseidon-amphibworks-is-combining-cars-boats-and-hydrofoils-for-the-next-new-mode-of-transportation/ (Accessed November 10, 2023).

[40] *Poseidon AmphibWorks*. Available: https://poseidonamphibworks.com/ (Accessed November 10, 2023).

[41] L. Blain. *Amphibious electric 3-wheeler will drive, float and hydrofoil*. Available: https://newatlas.com/marine/poseidon-amphibworks/ (Accessed November 10, 2023).

[42] *CT&T's multi-amphibious electric car cruises through water*. Available: https://www.fastcompany.com/1515698/ctts-multi-amphibious-electric-car-cruises-through-water (Accessed November 10, 2023).

[43] A. Haxhiu, A. Abdelhakim, S. Kanerva, and J. Bogen, "Electric power integration schemes of the hybrid fuel cells and batteries-fed marine vessels—An overview," *IEEE Transactions on Transportation Electrification*, vol. 8, no. 2, pp. 1885–1905, 2021.

[44] *MF Ampere – The world's first all-electric car ferry*. Available: https://corvusenergy.com/projects/mf-ampere/ (Accessed November 01, 2023).

[45] *Ferry system electrification*. Available: https://wsdot.wa.gov/construction-planning/major-projects/ferry-system-electrification (Accessed November 05, 2023).

[46] *World's first hybrid expedition ship to be christened in Antarctica*. Available: https://www.hurtigruten.com/us/press-releases/2019/worlds-first-hybrid-expedition-ship-to-be-christened-in-antarctica/ (Accessed November 10, 2023).

[47] A. Skredderberget. The first ever zero emission, autonomous ship. Available: https://www.yara.com/knowledge-grows/game-changer-for-the-environment/ (Accessed November 10, 2023).

[48] *Silent yachts*. Available: https://www.silent-yachts.com/silent60/ (Accessed November 05, 2023).

[49] *Oceanco unveils their new and innovative 115-metre Tuhura*. Available: https://www.theextravagant.com/lifestyle/yachts/oceanco-unveils-innovative/ (Accessed November 05, 2023).

[50] *CRP Azipod® is first choice for visionary superyacht Tuhura*. Available: https://new.abb.com/news/detail/3895/crp-azipodr-is-first-choice-for-visionary-superyacht-tuhura (Accessed November 05, 2023).

[51] J. Yusuf, A. J. Hasan, and S. Ula, "Impacts of plug-in electric vehicles on a distribution level microgrid," in *2019 North American Power Symposium (NAPS)*, IEEE, pp. 1–6, 2019.

[52] C. Schuss, B. Eichberger, and T. Fabritius, "An advanced PV simulation model for electric vehicles with photovoltaic installations," in *2021 IEEE International Instrumentation and Measurement Technology Conference (I2MTC)*, IEEE, pp. 1–6, 2021.

[53] A. Łebkowski, "Evaluation of the possibility of using hybrid electric-propulsion systems for inland barges," *TransNav: International Journal on Marine Navigation and Safety of Sea Transportation*, vol. 12, no. 2, pp. 261–269, 2018.

[54] J. Vavrin, "Power and energy considerations at forward operating bases (FOBs)," in *NDIA Environment, Energy Security & Sustainability (E2S2) Symposium & Exhibition*, Denver, CO, 2010.

[55] E. Catenaro, D. M. Rizzo, and S. Onori, "Framework for energy storage selection to design the next generation of electrified military vehicles," *Energy*, vol. 231, p. 120695, 2021.

[56] IEA, *Global Outlook 2023, Catching up with Climate Ambitions*. International Energy Agency, Paris, France, 2023.

[57] Y.-H. Luo, H.-x. Wei, L.-b Tang, *et al.*, "Nickel-rich and cobalt-free layered oxide cathode materials for lithium ion batteries," *Energy Storage Materials*, vol. 50, pp. 274–307, 2022.

[58] M. U. Mutarraf, Y. Guan, L. Xu, C.-L. Su, J. C. Vasquez, and J. M. Guerrero, "Electric cars, ships, and their charging infrastructure—A comprehensive review," *Sustainable Energy Technologies and Assessments*, vol. 52, p. 102177, 2022.

[59] L. Li, X. Cao, and S. Zhang, "Shared energy storage system for prosumers in a community: Investment decision, economic operation, and benefits allocation under a cost-effective way," *Journal of Energy Storage*, vol. 50, p. 104710, 2022.

[60] K. Khavarian and K. M. Kockelman, "Life-cycle analysis of electric vertical take-off and landing vehicles," *Transportation Planning and Technology*, vol. 47, no. 8, pp. 1227–1242, 2024.

[61] M. Manikandan, E. Vaidya, and R. S. Pant, "Design and analysis of hybrid electric multi-lobed airship for cargo transportation," *Sustainable Energy Technologies and Assessments*, vol. 51, p. 101892, 2022.

[62] F. Razi and I. Dincer, "A review of the current state, challenges, opportunities and future directions for implementation of sustainable electric vehicle infrastructure in Canada," *Journal of Energy Storage*, vol. 56, p. 106048, 2022.

[63] A. C. Ruoso and J. L. D. Ribeiro, "An assessment of barriers and solutions for the deployment of electric vehicles in the Brazilian market," *Transport Policy*, vol. 127, pp. 218–229, 2022.

[64] A. Pamidimukkala, S. Kermanshachi, J. M. Rosenberger, and G. Hladik, "Evaluation of barriers to electric vehicle adoption: A study of technological, environmental, financial, and infrastructure factors," *Transportation Research Interdisciplinary Perspectives*, vol. 22, p. 100962, 2023.

[65] H. Li, W. Zhang, X. Yang, *et al.*, "State of charge estimation for lithium-ion battery using an electrochemical model based on electrical double layer effect," *Electrochimica Acta*, vol. 326, p. 134966, 2019.

[66] Z. Zeng, J. Cheng, Y. Li, *et al.*, "Composite cathode for all-solid-state lithium batteries: Progress and perspective," *Materials Today Physics*, vol. 32, p. 101009, 2023.

[67] M. H. Hossain, M. A. Chowdhury, N. Hossain, M. A. Islam, and M. H. Mobarak, "Advances of lithium-ion batteries anode materials—A review," *Chemical Engineering Journal Advances*, vol. 16, p. 100569, 2023.

[68] M. Ulldemolins, F. Le Cras, and B. Pecquenard, "Memory effect highlighting in silicon anodes for high energy density lithium-ion batteries," *Electrochemistry Communications*, vol. 27, pp. 22–25, 2013.

[69] Y. Si, X. Liu, T. Wang, *et al.*, "State-of-the-art review and future trends of development of tidal current energy converters in China," *Renewable and Sustainable Energy Reviews*, vol. 167, p. 112720, 2022.

[70] S. M. Katalenich and M. Z. Jacobson, "Toward battery electric and hydrogen fuel cell military vehicles for land, air, and sea," *Energy*, vol. 254, p. 124355, 2022.

[71] C. Wu, D. Zhou, X. Lin, Q. Sui, F. Wei, and Z. Li, "A novel energy cooperation framework for multi-island microgrids based on marine mobile energy storage systems," *Energy*, vol. 252, p. 124060, 2022.

Chapter 4

AI applications for electrical vehicles integration: opportunities and challenges

Samuel Raafat Fahim[1], *Shady H.E. Abdel Aleem*[2] *and Hany M. Hasanien*[3]

4.1 Introduction

Nowadays, there is a tremendous increase in the global awareness with the global warming bad effects. This is aiming to the harsh weather conditions and elevated ambient temperature. One of the prominent solutions is the fleet electrification that arose as a revolutionary concept. This will benefit both the onshore and offshore fleets. For onshore transportation, the adoption of electric vehicles (EV) become a paramount. They took the center stage in the onshore transportation industry. EVs offers sustainable and cleaner alternative compared to traditional petrol-powered vehicles. Besides the opportunities offered by EVs, there are challenges that must be handled [1].

The opportunities of EVs lie in using electricity as a total driving power; or sometimes along with petrol-powered engines in hybrid EVs version. As a consequence, EVs can reduce the carbon dioxide emissions. Leading to a greener and more stable climate with enhanced air quality. Along with the previous, the EVs provides a lower operating costs compared to traditional vehicles. With minimal number of moving parts, comes another opportunity for lowering the required maintenance [2]. Last but not least, the operation of EVs are quieter, smoother, and enhance both the driving comfort and satisfaction.

However, in the previously mentioned opportunities, EVs have a fair share of undeniable challenges [3]. One of the limitations is the driving range with regard to their traditional counterparts. In addition, the availability of the charging stations plays a crucial limitation of spreading the EVs usage. Finally, the initial cost of EVs deters the switch from traditional vehicles to EVs. However, the costs of EVs are decreasing but in general they are more expensive than traditional ones [4].

[1]Independent Researcher, Electrical Power and Machines Department, Faculty of Engineering, Ain Shams University, Cairo, Egypt
[2]Department of Electrical Engineering, Institute of Aviation Engineering and Technology, Giza, Egypt
[3]Electrical Power and Machines Department, Faculty of Engineering, Ain Shams University, Cairo, Egypt

To oppose the aforementioned challenges, there are several solutions that can be offered. First, investing in enhancing the battery and fuel cell (FC) technologies. Their advancement can increase the EVs range capabilities and alleviate the range anxiety. Second, spreading stations with fast charging technologies. This reduces the down time of the EVs impacting the practicality of using EVs. Third, the governments, organizations, and private companies should offer incentives and subsidies to societies. This in return helps in offsetting the initial costs of EVs which makes EVs solutions, to both ordinary users and fleet operators, a more appealing option. Fourth and final, the integration of the EVs and DISCOs (distribution grid companies) is called vehicle-to-grid (V2G). This provides an opportunity to EVS owners to participate in grid demand response; thus, the owners and operators can monetize their EVS assets [5].

Introducing AI in integrating the EVs with the electrification of the fleet exploits the benefits and mitigates the challenges at the same time. Adopting AI algorithms can provide optimized routing for large fleets and efficient charging strategies, thus elevating the overall operation efficiency of EVs. For the individual EV owners, AI algorithms can analyze real-time traffic data. Additionally, it can provide energy consumption and charging station status. All of this contributes in decreasing the downtime and maximizing the EVs operation. Besides, AI algorithms enable predictive maintenance schedules, which enables to monitor the problems before they escalate. The AI-powered services ensures cost-effective charging regarding the grid electrical demand and pricing, furthermore minimizing congestions in electrical grid [6].

This chapter will cover aforementioned points in detail in the upcoming sections, which is in a way both simple for the non-specialists and in depth for the expert readers.

4.2 Classification of EVs

The vehicles were originally powered by batteries since the early 20th century. However, due to the need for more speed and range, and the limitation for the batteries sizes, the industrial revolution offered internal combustion engines (ICE). Then after almost 100 years, and with the advancement of batteries, the EVs idea came on top again.

First, there were hybrid vehicles that used both the ICE along with electric sources (batteries or FC later). Then came the all-electric vehicle.

In this section, a brief comprehensive demonstration for the various types of the vehicles is provided, mainly concentrating on EVs and its variations. In Figure 4.1, a flow chart lists all vehicles types ranging from all electric vehicles (AEVs) that use either batteries or FCs along with the hybrid vehicles that use both the ICE and electric sources [7–9].

4.2.1 Hybrid electric vehicle

Hybrid electric vehicles (HEVs) rely on a combination of traditional ICE and an electric motor (EM) for propulsion. Unlike plug-in hybrid electric vehicles

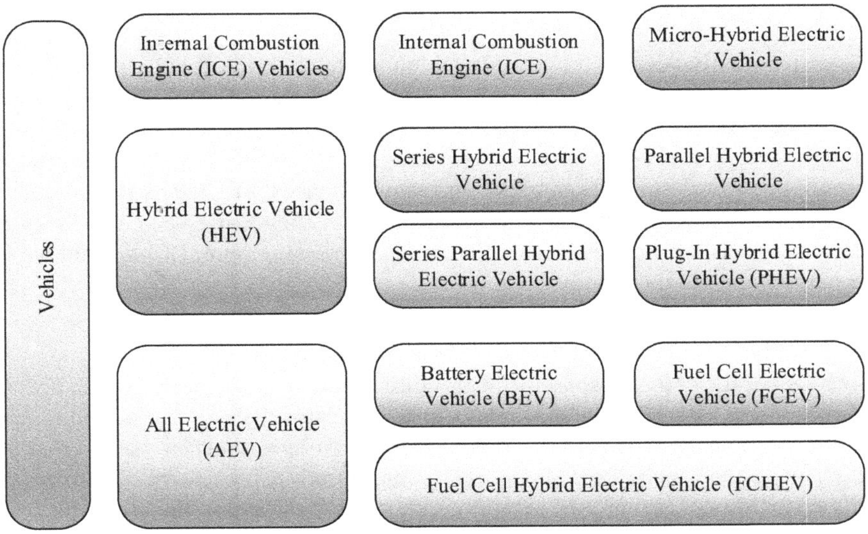

Figure 4.1 List of all vehicles types

(PHEVs), HEVs do not have the capability to be connected to the electrical grid. Instead, the battery that supplies power to the EM is recharged through the energy generated by the vehicle's combustion engine. In modern HEV models, the batteries can also be recharged through regenerative braking, where kinetic energy is converted into electrical energy. For instance, the fourth generation of the Toyota Prius, in its hybrid variant, was equipped with a 1.3 kWh battery that theoretically provided a range of up to 25 kilometers when operating in its all-electric mode [10].

4.2.2 Series hybrid electric vehicle

A series hybrid is a type of hybrid system in which the tractive force is solely provided by EMs, while an ICE is coupled with a generator to produce electricity for powering the EMs, propelling the vehicle. In series hybrids, the power rating of the EM is higher compared to parallel hybrids. Sometimes, series hybrids are referred to as range-extended EVs. The most significant advantage of this concept is that the ICE can operate at its highest efficiency since the engine and vehicle RPM are independent, resulting in reduced fuel consumption [11,12].

However, this system implies larger electrical machines, a substantial battery pack, and a battery charging setup via an electric generator. The essential power electronic components in series HEVs include rectifiers (which convert the AC output from the alternator to DC for battery charging), a DC-DC converter, and an inverter (for converting DC to AC) for motor propulsion. Series HEVs are known for their low maintenance requirements and extended lifespan, making them well suited for urban driving conditions. The Chevrolet Volt serves as an example of a series HEV [13].

4.2.3 Parallel hybrid electric vehicle

Within parallel HEVs, the tractive force required to drive the vehicle's wheels can be derived from both the ICE and the EM, which operate in parallel during acceleration and braking. The ICE and EM are directly connected to the vehicle's wheels through a mechanical transmission system, allowing for the potential reduction in the ratings of both the ICE and EM. In many instances, the ICE is the dominant power source and operates continuously. One of the pioneering vehicles to feature this system was the Renault Scenic. Examples of parallel hybrid EVs available in the market include the Honda Civic Hybrid [14,15].

4.2.4 Series-parallel hybrid electric vehicle

In series-parallel HEVs, apart from the direct coupling of the ICE and EM in parallel mode, the ICE is also linked to a generator. This generator produces electricity to charge the battery via a power electronics interface (series mode), which in turn propels the vehicle. This configuration incorporates two power delivery pathways (series and parallel), effectively combining the characteristics of both series and parallel hybrids, earning it the designation of a dual-mode HEV [16,17].

The series-parallel hybrid is alternatively referred to as a power-split hybrid, where a mechanism splits power between the ICE and EM. This power-splitting device can either be mechanical, like a planetary gearbox, or electrical, involving an electric power splitter. For instance, a planetary gearbox is utilized in vehicles such as the Toyota Prius or Lexus RX400h, both of which are under the Toyota umbrella.

In the series pathway, a connection exists between the ICE and the electric generator. The primary advantage of this system is its versatility, capable of operating in both series and parallel modes [18].

4.2.5 Plug-in hybrid electric vehicle

Plug-in hybrid electric vehicles (PHEVs) represent an innovative and forthcoming technology with implications for both the transportation and power sectors. PHEVs are vehicles equipped with batteries or another form of energy storage that can be recharged by connecting to an external source of electrical power (excluding electrical grid). They serve as a compromise between traditional gasoline-powered vehicles and fully electric ones, operating primarily on electric power while also having the capability to switch to the ICE when the battery is depleted. Notable advantages of PHEVs include reduced fuel consumption and lower emissions, which, however, need to be weighed against the relatively high cost of the battery components. Examples of PHEVs include models like the Chevy Volt and Toyota Prius [19–21].

For instance, the Mitsubishi Outlander PHEV is equipped with a 12 kWh battery, allowing it to cover approximately 50 kilometers solely on electric power. Nevertheless, it is important to note that the actual fuel consumption of PHEVs may exceed the figures provided by car manufacturers [22].

4.2.6 Battery electric vehicle

All vehicles operate entirely on electric power. Battery electric vehicles (BEVs) lack an ICE and do not rely on any form of liquid fuel. It offers the advantage of emitting no harmful pollutants into the environment, but it comes with limitations related to intermittent recharging and range coverage. The straightforward powertrain design, minimal energy conversion losses (from electrical to mechanical), reduced vehicle weight, and environmentally friendly attributes are some of the positive aspects of BEVs [23,24].

They typically employ substantial battery packs to provide the vehicle with a reasonable driving range. A typical BEV can cover distances between 160 and 250 kilometers on a single charge, although some can extend to 500 kilometers. The Nissan Leaf is an example of such a vehicle, being 100% electric and equipped with a 62 kWh battery, enabling users to drive up to 360 kilometers on a single charge [25]. However, there are a few drawbacks, such as the need for a high-torque traction motor, relatively long recharging times, and the necessity for effective battery management.

4.2.7 Fuel cell electric vehicle

These vehicles are equipped with an EM that utilizes a combination of compressed hydrogen and oxygen from the air, resulting in water as the sole byproduct of this process. While these vehicles are often touted as having "zero emissions," it's important to note that, despite the availability of green hydrogen, a significant portion of the hydrogen used is still derived from natural gas. The Hyundai Nexo FCEV is an example of such vehicles, boasting a travel range of 650 kilometers without the need for refueling [26–28].

Fuel cell-based electric vehicles (FCEVs) exclusively rely on FCs for power. FCEVs have gained increasing popularity due to their high efficiency and minimal FC emissions. Among the various types of commercially accessible FCs, the proton exchange membrane fuel cell (PEMFC) is the preferred choice. The powertrain architecture of FCEVs mirrors that of BEVs. However, the primary drawback of FCEVs lies in the elevated cost of the fuel, namely hydrogen (H2). Examples of FC-based vehicles include the Honda FCX Clarity, Hyundai Tucson, and Toyota Mirai [29].

4.2.8 Fuel cell hybrid electric vehicle

A hybrid system that combines both a battery and a FC can be employed to provide power for the vehicle. By integrating an energy storage system (ESS) and appropriate power electronic converters, a new vehicle configuration is established, known as a FC-based hybrid electric vehicle (FCHEV). The FCHEV is characterized by its eco-friendly, as it operates without emitting harmful pollutants. In this configuration, the FC serves as the primary power source, generating electricity to supply the EM for vehicle propulsion. Both the battery and the Fuel Cell contribute to electricity generation, making their hybrid combination fitting within the category of AEVs [30].

In conclusion, and by comparing all the aforementioned EV types, the FCHEV is a potential technology to be explored emphatically in future. The FCHEV can play a significant role in the all-future HEV. As the main issue that faces the fuel cells-based EVs is the related to the availability, cost, production, and storage of hydrogen H_2. If these issues could be resolved, this will give a boost to the FCHEVs in general as the H_2 offers zero emission, high efficiency, and high driving range.

4.3 New storage technologies for EVs

Batteries stand out as a crucial component in EVs, holding significant importance as one of the most expensive elements contributing to the overall vehicle cost. Batteries play a direct role in shaping the performance of EVs. Advances in durability, charging densities, and the efficiency of charge and discharge processes have prompted the exploration of various resources for developing technologies that surpass the currently dominant lithium-ion batteries extensively utilized in vehicles.

Enhancements in battery technologies have the potential to expedite the success of EVs and contribute significantly to the widespread adoption of these vehicles globally. Currently, extensive research is underway to explore new technologies and components in this field.

In this section, only the new energy storage technologies will be presented. The reader can get back to [31], for a comprehensive details and modeling for various types of energy storage in general and batteries in specific. Some of the new innovations in the batteries field include the following:

4.3.1 Lithium iron phosphate ($LiFePO_4$)

This specific battery type poses an energy density of around 220 Wh/L, impressive durability (charging cycles ranges from 2,000 and 10,000 cycles), and the ability to withstand high temperatures. Despite ongoing testing of this battery type in EVs [32], it remains in the early stages of research and development. Researchers at MIT have made significant strides in reducing its weight and have successfully created a prototype cell that can achieve a complete charge in just 10–20 seconds, remarkably shorter when compared to the standard 6 minutes required for conventional battery cells [33].

4.3.2 Lithium metal

In these batteries, the graphite anode is substituted with a thin layer of lithium metal. This particular battery variant has the capability to store twice the power compared to traditional lithium batteries [34]. Lithium-metal batteries exhibit a notable Coulombic efficiency, surpassing 99.1%. They have a life span of more than 6,000 charging cycles. Even after 1,000 cycles, they maintain an average Coulombic efficiency of 98.4% [35].

4.3.3 Lithium air

This type of battery requires a continuous supply of oxygen to facilitate the reaction with lithium. Although first proposed in the 1970s, it is only recently that efforts have been made to develop and enhance these batteries. Anticipated advancements suggest that their specific energy could reach approximately 12 kWh/kg, nearly 45 times the current capacity of lithium batteries. This would position them at a comparable level to conventional fuel [36].

4.3.4 Magnesium-ion (Mg-ion)

These batteries shift from using lithium to magnesium, achieving the ability to store more than double the charge while enhancing overall stability. It is anticipated that this battery type could attain an energy density of 6.2 kWh/L, marking an increase of 8.5 times compared to the best-performing lithium batteries, which presently reach up to 0.735 kWh/L [37].

4.3.5 Aluminum air

Batteries crafted with this technology generate electricity through the interaction of oxygen with aluminum. The key benefit of this battery type is its capacity to achieve remarkably high energy densities, reaching up to 6.2 kWh/L [38], enabling substantial driving ranges of up to 1,600 km [39]. The cost of these batteries is on a downward trend, for about 300 € for each kWh [40]. Moreover, a notable advantage is their recyclability.

4.3.6 Sodium air (Na_2O_2)

This battery has an energy density of 4.5 kWh/L [41]. In the context of EVs, this battery variant has the potential to increase the range by at least 13 times compared to current lithium batteries [42]. A significant advantage of these batteries lies in the abundance of sodium, which ranks as the sixth most plentiful element on our planet [43].

4.3.7 Graphene

Graphene is a material that is formed by pure carbon, which has a high thermal conductivity and it is extremely light (a one square meter blade weighs 0.77 mg) [44]. A notable advantage of batteries based on graphene is their minimal heat generation, facilitating rapid or ultra-fast charging without substantial power losses due to heat. A graphene battery was created that was used in vehicle, with 900 HP, and was able to travel for 800 km [45]. In a high-power plug, and in only 5 minutes, the battery could be fully charged. This battery technology is still under further developments. However, there are prototypes from graphene batteries with power of 1 kWh/kg. In near future, it is expected to achieve 6.4 kWh/kg [46].

4.3.8 Emerging battery technologies

Along with the aforementioned technologies in details, it is worth mentioning some of the emerging technologies. Among these technologies are three categories, the first category contains:

- Dual-ion battery (DIB) [47]
- Dual graphite/carbon battery
- Aluminum-graphite DIB [48]

While the second category contains dual carbon battery (DCB) [49], the third and final category contains dual carbon-based potassium dual ion [50].

All poses promising large energy density which can enable them to replace the ordinary EV batteries due to their remarkable performance.

However, apart from the above-mentioned advantages, they are associated with the following disadvantages:

- Self-discharge,
- Poor safety,
- Low Coulombic efficiency, and
- Low energy density.

The ongoing research considers the above-mentioned disadvantages and try to find the combination and new electrolytes that enhances the performance of both DIB and DCB batteries [51].

4.4 Challenges for EVs integration

In this section, the challenges that face the EVs integration to the electrical grid are presented.

There are several technological and infrastructure challenges surrounding widespread EV adoption [52]. Here are some of the challenges:

- *EV prices*: EVs need price parity with gas/petrol vehicles (ICE) to attract more consumers [53].
- *Battery issues*: battery efficiency decreases at high speeds and loads, so improving power and performance is key. This impacts on the EVs dynamic performance and driving range, especially the AEV types [54,55]. In addition, the ESS takes up a significant space and weight [56–58]. Setting global sourcing and pricing volatility for battery materials like lithium, nickel and cobalt, and graphite impact total costs [59]. Not to forget that the proper disposal and recycling of battery and fuel cell components after use presents environmental concerns due to hazardous materials [60].
- *Privacy and security*: quality, safety, and cybersecurity standards must meet consumer expectations. As the integration of EVs implies the gathering and examination of delicate user information, including charging habits and locations. Safeguarding the privacy and security of this data poses a significant challenge for AI systems, demanding strong encryption, anonymization methods, and secure protocols for data management [61].
- *Reliability:* key drivetrain parts like power converters, batteries, EMs, and fuel cell stacks need improvement to satisfy consumers [62].
- *Sustainability:* another potential hurdle for the EV industry is sustainability. While EVs have been tipped as more environmentally friendly than ICE vehicles, apprehensions have arisen regarding their sustainability. This is primarily attributed to the varied raw materials employed in EV battery production, and their availability. Components such as nickel, cobalt, and lithium are

utilized in the manufacturing of EVs, and the mining methods for these materials are largely deemed unsustainable [63].

- *Grid integration:* Cities experiencing substantial growth in EVs powered by renewable energy face a notable challenge in upgrading their existing infrastructure. The difficulty often lies in enhancing the flexibility of loads. Opportunities arise for EVs to contribute to local or bulk flexibility through V1G or V2G, thereby promoting increased adoption of renewable energy. Additionally, ongoing research by various industry and academic institutions aims to introduce innovative grid integration frameworks, facilitating the widespread adoption of EVs. This additional electricity demand as a consequence to the spread of charging stations risks overburdening grids if it is left without proper management [64].
- *Charging infrastructure*: significant refurbishment of current infrastructures, including residential buildings, community parking, and commercial structures, is essential to establish a robust EV charging ecosystem. Integrating renewable energy, such as solar panels, into these existing infrastructures can alleviate the strain on the grid and diminish the constant reliance on it. This approach not only aids in reducing the need for frequent visits to nearby charging stations but also allows surplus electricity generated by the renewable source to be exported to the grid through a grid-tie inverter. On the other hand, building out a reliable, affordable public charging network presents difficulties in determining costs and finding suitable urban land [65].
- *Governmental policies*: dynamic, coherent government policies are still needed to incentivize the entire EV ecosystem through manufacturing, technology, and infrastructure support [66].
- *Scalability*: as the number of EVs increases, this adds tons of large-scale data that express all the information about the integrated EVs. The management of these data, including the charging patterns and grid management, adds a complex challenge that the grid operators face [67].
- *Uncertainty and variability*: the charging patterns of EVs and the conditions of the grid can be unpredictable and subject to significant fluctuations. Effectively managing this uncertainty and variability poses a substantial challenge in making precise predictions and optimizing charging strategies accordingly [68].
- *Interoperability*: the integration EVs with electrical grids necessitate interoperability among diverse stakeholders, including EV manufacturers, charging infrastructure providers, grid operators, and participants in the energy market. The challenge here lies in developing adequate algorithms that can integrate and communicate with different platforms and systems, which is a challenge [69].
- *Optimization tradeoffs*: with the growing number of integrated EVs, it is of crucial importance to strike a balance between optimizing individual EV charging preferences and overall grid efficiency. Balancing user preferences, grid constraints, energy costs, and environmental factors requires sophisticated optimization techniques that can handle complex tradeoffs [70].
- *Grid congestion management*: this is one of serious issues imposed by EVs integration. The increased demand for EV charging can lead to grid congestion,

especially during peak periods. The management lies in reliving the grid congestion by optimizing charging schedules, load balancing, and coordinating with grid operators to ensure grid stability and prevent overloading [71].

- *User behavior and preferences*: understanding and accommodating user behavior and preferences is crucial for successful EV integration. The following factors shall be considered such as individual charging patterns, preferred charging times, travel needs, and user feedback to provide personalized and satisfactory charging experiences [72].
- *Energy market integration*: EV integration involves interactions with energy markets such as time-of-use pricing and demand response programs. This includes incorporate market dynamics, pricing signals, and market participation mechanisms to optimize EV charging strategies and enable efficient energy market integration [73].
- *Long-term planning and investment*: integrating EVs into electrical grids requires long-term planning and investment in infrastructure such as charging stations, grid upgrades, and renewable energy generation. Challenges include long-term planning by analyzing data, predicting future EV adoption trends, and optimizing infrastructure deployment strategies [74].
- *Education and user adoption*: promoting user awareness, education, and adoption of EVs and smart charging practices is a challenge that impacts the successful integration of EVs into electrical grids [75].

Addressing these issues will help smooth the transition to more electric transportation alternatives.

4.5 Opportunities of EVs integration

Although there are challenges that face the EVs integration into the electrical grid, there are opportunities that can be gained. In this section, a short list of these opportunities is presented.

- *Demand response:* EVs can be used as flexible energy resources to participate in demand response programs, responding to grid conditions and market signals to adjust their charging patterns. To harness the benefits of this opportunity, the participation of the EVs shall be optimized, thus enabling efficient grid management and cost savings for EV owners [76].
- *Renewable energy integration:* EVs can be charged using renewable energy sources, such as solar and wind power, enabling sustainable and carbon-neutral transportation. For maximizing the outcome of this integration, the following factors shall be considered: weather conditions, energy storage, and grid constraints [77].
- *Grid balancing:* EVs can be used to balance the electrical grid by providing energy storage and load-shifting capabilities. By considering grid stability, energy costs, and user preferences, this opportunity could be exploited effectively [78].

- *Vehicle-to-grid (V2G) integration:* V2G technology allows EVs to not only consume energy from the grid but also supply energy back to the grid. This integration shall be optimized with regard to both the grid and vehicle conditions. Thus EVs integration can be fruitful to energy markets; for example, providing ancillary services and contributes to grid stability [79].
- *Grid resiliency:* The integration of EVs into electrical grids can improve grid resiliency by providing backup power and load-shifting capabilities during emergencies or outages. Optimized integration of EVs can enhance grid resiliency, along with considering grid stability, user needs, and energy costs [80].
- *Grid services provision:* EVs can provide various grid services such as frequency regulation and voltage support. It is worth mentioning that by aggregating and coordinating the capabilities of multiple EVs, this contributes to grid stability and reliability [81].
- *EVs shared mobility:* Shared mobility services, such as ride-sharing and car-sharing, are increasingly adopting EVs. This can be possible by managing charging schedules, coordinating vehicle availability, and optimizing fleet operations [82].
- *Grid-interactive buildings:* EVs can be integrated with smart buildings to support grid interactions such as demand response programs and building-to-grid (B2G) interactions. To maximize energy efficiency and grid benefits, the following factors shall be optimized and coordinated: EV charging, building energy management systems, and grid operations [83].
- *EV energy trading:* The integration of EVs into the electrical network presents opportunities for peer-to-peer energy trading and demand response programs. By introducing organizing algorithms to EV energy trading, this enables EV owners to sell excess energy back to the grid or other users [84].
- *EV integration in smart cities:* The integration of EVs into the electrical network is a key component of smart city initiatives. This integration can provide efficient and sustainable transportation systems which in return can be extended to include integration with multiple smart cities [85].
- *EV integration in rural areas:* Rural areas often face unique challenges in terms of limited charging infrastructure and longer travel distances. However, this kind of integration needs tailoring solutions to make EV integration in rural areas beneficial. Those solutions might be optimizing charging infrastructure deployment, range prediction models, and community-based charging strategies [86].
- *Electrification of public transportation:* The electrification of public transportation systems, such as buses and taxis, can have a significant impact on reducing emissions and improving air quality. This trend shall provide solution to many aspects such as optimizing the charging infrastructure, scheduling, and routing of electric public transportation vehicles. Also, keep in consideration the passenger demand, traffic conditions, and operational requirements [87].

4.6 AI application in EVs

AI application can diminish the problems that arise from the challenges apart from harnessing the benefits from the opportunities, both of which were aforementioned in this chapter. However, this section provides additional new AI applications for introducing EVs to the electrical grid.

- *Grid-aware routing:* AI algorithms can provide grid-aware routing for EVs, considering factors such as charging station availability, congestion management, energy prices, and grid constraints. This can optimize travel routes for EVs, ensuring efficient charging opportunities and minimizing range anxiety [88].
- *Predictive maintenance:* AI algorithms can analyze data from EVs and charging infrastructure to predict maintenance needs and optimize maintenance schedules. This can help prevent breakdowns, reduce downtime, and ensure the reliability of EV charging infrastructure [89].
- *Grid optimization:* AI algorithms can optimize the operation and management of electrical grids by leveraging the flexibility of EV charging. By coordinating EV charging patterns with grid conditions, AI can help balance supply and demand, reduce peak load, and improve overall grid efficiency [90].
- *Energy forecasting:* AI algorithms can forecast energy demand and supply by gathering and analyzing historical data, weather patterns, and user behavior to forecast. This can help grid operators and energy market participants make informed decisions regarding grid operations, renewable energy integration, and EV charging management [91].
- *Optimal charging and discharging strategies:* While AI algorithms have been developed to optimize charging and discharging of EVs, there will be always a need for further research to explore more advanced and efficient strategies. This includes considering factors such as grid constraints, renewable energy integration, user preferences, and dynamic pricing mechanisms to develop optimal charging and discharging schedules for EVs [92].
- *Policy and regulatory frameworks:* The integration of EVs into electrical networks requires supportive policy and regulatory frameworks. Research is needed to analyze the impact of different policy measures, such as incentives, regulations, and standards, on the successful integration of EVs. AI can be used to model and simulate the effects of various policy scenarios to guide decision-making processes [93].
- *Social and environmental impacts:* The widespread adoption of EVs has social and environmental implications that need to be studied. Research is needed to assess the social acceptance, equity considerations, and environmental impacts of EV integration. AI can be used to model and analyze these impacts, informing policy decisions and promoting sustainable EV integration [94].
- *Multi-objective optimization:* The integration of EVs into electrical networks involves multiple objectives such as minimizing energy costs, reducing emissions, and improving grid stability. AI-based algorithms can optimize multiple objectives simultaneously, considering the trade-offs and synergies between different objectives [95].
- *Data analytics and visualization:* The integration of EVs generates large amounts of data that need to be analyzed and visualized to inform decision-

making processes. Improving existing AI-based tools is needed for data analytics and visualization that can extract insights from data and provide actionable recommendations for EV integration [96].

- *Vehicle-to-home (V2H) integration:* V2H technology allows EVs to discharge power back to the home or the grid during peak demand or power outages. Investing research effort in AI-based solutions is of crucial importance to better address and explore the potentials of V2H integration, including optimal power flow management, load balancing, and coordination with renewable energy sources [97].
- *Grid integration of autonomous EV fleets:* the emergence of autonomous EV fleets presents unique challenges for grid integration. Research is needed to develop AI algorithms that can optimize the charging and routing of these autonomous fleets, considering factors such as energy efficiency, charging infrastructure availability, and fleet management strategies [98].
- *Infrastructure planning:* planning the deployment of EV charging infrastructure requires careful consideration of factors such as location selection, capacity planning, and cost optimization. The AI-based optimization techniques provide a promising assistance in the planning and design of EV infrastructure to ensure efficient utilization and scalability [99].
- *Cybersecurity:* AI can also enhance cybersecurity measures in the context of EV integration. As EVs become more connected and communicate with the grid, they become potential targets for cyber-attacks. AI algorithms can analyze network traffic, identify anomalies, and detect potential threats in real time. By leveraging AI's ability to learn and adapt to new attack vectors, it can strengthen the security of EVs and the overall grid infrastructure [100].
- *Dynamic wireless charging:* Dynamic wireless charging technology allows EVs to charge while on the move, eliminating the need for frequent stops at charging stations. There are ongoing studies focusing on developing AI algorithms that optimize dynamic wireless charging systems, considering factors such as vehicle speed, traffic conditions, energy efficiency, and charging infrastructure deployment [101].

4.7 AI techniques for handling EV integration

The AI techniques have well-founded basics and categories, from fuzzy logic (FL), artificial neural networks (ANN), to minimum seeking algorithms, and finally expert systems. However, the innovation in these concepts is ongoing strongly. Also, the integration between these techniques opens new possibilities and promotes renovated solutions. By leveraging these techniques, we can ensure a smooth and successful integration of EVs into the grid, creating a more sustainable and efficient energy system.

This section will list a sample of the AI techniques and its uses in the field of EV integration into electrical grids.

- *Machine learning:* A prominent AI technique which can be used to analyze historical data on charging patterns and predict future demand. This information can be used to optimize charging schedules, ensuring that EVs are charged at times when demand on the grid is low and avoiding overloading the grid during peak periods [102].

- *Reinforcement learning:* this technique involves training an AI system to make decisions based on a reward system. In the context of integrating EVs into the grid, reinforcement learning can be used to optimize charging schedules by incentivizing EV owners to charge their vehicles at specific times. For example, EV owners could be rewarded for charging their vehicles during off-peak periods, reducing the load on the grid during peak demand periods [103].
- *Data analytics:* utilizing this AI technique, one can identify patterns and correlations that can help optimize grid management. This data-driven approach can provide insights into charging behaviors, grid capacity, and demand fluctuations, enabling better decision-making and resource allocation. The data sources could be from various sources. Among which are the EV charging stations, grid sensors, weather forecasts, and output from other AI algorithms [104].
- *Optimization algorithms:* these algorithms can, but not limited to, analyze various parameters such as electricity prices, grid constraints, and EV charging requirements to determine the most efficient and cost-effective charging schedules by considering factors such as grid capacity, renewable energy generation, and EV owner preferences, AI optimization algorithms can balance the charging load across different time periods and locations, reducing the strain on the grid during peak demand [105].
- *Multi-agent systems:* Multi-agent systems involve multiple intelligent agents interacting with each other to achieve common goals. The daily published literature provides novel techniques/enhancements to the multi-agent algorithms in general and to EV integration in specific. These techniques enable coordination and cooperation among EVs, charging infrastructure, grid operators, and other stakeholders to optimize charging strategies and grid operations [106].
- *Edge computing and AI:* this involves processing data closer to the source, thereby enhancing the efficiency and responsiveness of AI applications for EV integration. The ongoing research can explore the integration of edge computing and AI techniques to enable real-time decision-making, reduce latency, and improve privacy in EV charging and grid interactions [107].
- *Explainable AI:* As AI algorithms become more complex, it becomes crucial to ensure transparency and interpretability. Future research can focus on developing explainable AI techniques for EV integration, allowing stakeholders to understand the decision-making process of AI models and build trust in their outcomes [108].
- *Federated learning:* Federated learning is a distributed machine learning technique that allows multiple parties to collaboratively train a model without sharing their data along with efficient learning from diverse datasets and reduced communication overhead. However, the following aspects shall be considered, such as model security, precise consideration to the heterogeneous nature of the EV fleets, and managing communication latency [109].

4.8 Future research trends

The integration of EVs to the electrical grid is an inevitable event that is happening and gaining a growing market share. This integration will impact all the aspects listed in the chapter ranging from challenges to opportunities. These aspects cover all the levels from the single consumer to the shareholders ending with the operators of the electrical grids, either entities or governments.

Here comes the importance of applying AI algorithms that are playing and will continue in affecting all aspects of EVs integration. AI algorithms will enable the harnessing of the benefits and maximize the opportunities as well as mitigate the challenges. The importance of AI algorithms rises from the growing complexity of aspects related to the EVs integration.

The application of AI algorithms will not only be concerned and limited to the technical issues listed in the chapter but also associated with the legal and social aspects. AI will also handle the security issues that are related to the IoT and big data.

The future research will not be limited to the AI applications, but will also be related to the development of the AI techniques itself, thus to find more eligible solutions that absorb the complex nature and the growing size of the data that needs to be handled along with the interleaving relations between inputs and outputs.

4.9 Conclusions

This chapter was concerned with presenting the different aspects of the EVs integration to electrical grids with a specific focus on the AI applications to the coexisting challenges and opportunities and a demonstration to every aspect ranging from EV types, challenges, opportunities, and AI applications.

It was clear either the challenges or the opportunities were of complex nature. This contains many aspects that need careful addressing. Thus, the decision-making, management, and optimization are more realistic along with the social and behavioral issues that have non-technical nature. Not forgetting the security issues that arise from sharing and connecting a lot of parties to a single grid.

All of this implies the need for the precise selection of a proper AI tool as well as the consideration of the associated aspects, thus reaching an appealing solution to the underlying problem.

References

[1] J. Lu, R. Shan, N. Kittner, W. Hu, and N. Zhang, "Emission reductions from heavy-duty freight electrification aided by smart fleet management", *Transportation Research Part D: Transport and Environment*, vol. 121, pp. 103846, 2023.

[2] A. G. Olabi, M. A. Abdelkareem, T. Wilberforce, *et al.*, "Strength, weakness, opportunities, and threats (SWOT) analysis of fuel cells in electric vehicles", *International Journal of Hydrogen Energy*, vol. 48, no. 60, pp. 23185–23211, 2023.

[3] K. Deepak, M. A. Frikha, Y. Benômar, *et al.*, "In-wheel motor drive systems for electric vehicles: State of the art, challenges, and future trends", *Energies*, vol. 16, no. 7, pp. 3121, 2023.

[4] A. J. Ernst, C. R. Wade, D. G. Dyer, D. K. Danquah, and O. S. Tomomewo, "The Impact of the Electric Vehicle Transition on US Grid Infrastructure and Energy Policy", *American Journal of Energy Research*, vol. 11, no. 1, pp. 1–4, 2023.

[5] M. Rahmani-Andebili, "Sensitivity analyses for optimal charging management of electric vehicles in San Francisco", in: M. Rahmani-Andebili (eds.) (ed.), *Planning and Operation of Electric Vehicles in Smart Grids*, Springer, Cham, pp. 53–68, 2023.

[6] J. Li, M. S. Herdem, J. Nathwani, and J. Z. Wen, "Methods and applications for Artificial Intelligence, Big Data, Internet of Things, and Blockchain in smart energy management", *Energy and AI*, vol. 11, pp. 1–18, 2023.

[7] S. R. Salkuti, "Energy storage and electric vehicles: technology, operation, challenges, and cost-benefit analysis", *International Journal of Advanced Computer Science and Applications*, vol. 12, no. 4, pp. 40–45, 2021

[8] D. Kumar, R. K. Nema, and S. Gupta, "A comparative review on power conversion topologies and energy storage system for electric vehicles", *International Journal of Energy Research*, vol. 44, no. 10, pp. 7863–7885, 2020.

[9] J. A. Sanguesa, V. Torres-Sanz, P. Garrido, F. J. Martinez, and J. M. Marquez-Barja, "A review on electric vehicles: Technologies and challenges", *Smart Cities*, vol. 15, no. 4, pp. 372–404, 2021.

[10] S. Hirota, T. Kikuchi, and T. Katanoda, "*Development of the new 2.0 L hybrid system for Prius*", SAE Technical Paper, 2023.

[11] F. Slah, A. Mansour, M. Hajer, and B. Faouzi, "Analysis, modeling and implementation of an interleaved boost DC-DC converter for fuel cell used in electric vehicle", *International Journal of Hydrogen Energy*, vol. 42, no. 48, pp. 28852–28864, 2017.

[12] A. Garrigos, and F. Sobrino-Manzanares, "Interleaved multi-phase and multi-switch boost converter for fuel cell applications", *International Journal of Hydrogen Energy*, vol. 40, no. 26, pp. 8419–8432, 2015.

[13] S. E. de Lucena, "A survey on electric and hybrid electric vehicle technology", in: S. Soylu (eds.) (ed.), *Electric Vehicles: The Benefits and Barriers*, IntechOpen, London, pp. 1–21, 2011.

[14] A. M. Lulhe, and T. N. Date, "A technology review paper for drives used in electrical vehicle (EV) & hybrid electrical vehicles (HEV)", *2015 International Conference on Control, Instrumentation, Communication and Computational Technologies (ICCICCT)*, pp. 632–636, 2016.

[15] Z. Yuan, L. Teng, S. Fengchun, and H. Peng, "Comparative study of dynamic programming and Pontryagin's minimum principle on energy management for a parallel hybrid electric vehicle", *Energies*, vol.6, no. 4, pp. 2305–2318, 2013.

[16] A. K. Gupta, B. A. Landge, and B. Seth, "*Development of a parallel hybrid electric vehicle*", SAE Technical Paper, 2009.

[17] A. Pedram Pourhashemi, S. M. Ansarey Movahed, and M. Shariat Panahi, "Application of the fuel-optimal energy management in design study of a parallel hybrid electric vehicle", *Journal of Fuels*, vol. 2014, Article ID. 417172, pp. 1–12, 2014.
[18] J. K. Kim, E. S. Yim, C. H. Jeon, C. S. Jung, and B. H. Han, "Cold performance of various biodiesel fuel blends at low temperature", *International Journal of Automotive Technology*, vol. 13, no. 2, pp. 293–300, 2012.
[19] J. L. Torres, R. Gonzalez, A. Gimenez, and J. Lopez, "Energy management strategy for plug-in hybrid electric vehicles. A comparative study", *Applied Energy*, vol. 113, pp. 816–824, 2014.
[20] R. A. Waraich, M. D. Galus, C. Dobler, M. Balmer, G. Andersson, and K. W. Axhausen, "Plug-in hybrid electric vehicles and smart grids: investigations based on a microsimulation", *Transportation Research Part C: Emerging Technologies*, vol. 28, pp. 74–86, 2013.
[21] B. Lane, B. Shaffer, G. S. Samuelsen, "Plug-in fuel cell electric vehicles: A California case study", *International Journal of Hydrogen Energy*, vol. 42, no. 20, pp. 14294–14300, 2017.
[22] Mitsubishi Motors. Mitsubishi Outlander PHEV 2018. 2019. Available online: https://www.mitsubishicars.com/outlander-phev/2018/specifications (Accessed: Nov 2023).
[23] W. Li, R. Long, H. Chen, and J. Geng, "A review of factors influencing consumer intentions to adopt battery electric vehicles", *Renewable and Sustainable Energy Reviews*, vol. 78, pp. 318–328, 2017.
[24] A. Schuller, B. Dietz, C. M. Flath, and C. Weinhardt, "Charging strategies for battery electric vehicles: economic benchmark and V2G potential", *IEEE Transactions on Power Systems*, vol. 29, pp. 2014–2022, 2014.
[25] InsideEVs. Nissan Reveals LEAF e-Plus: 62 kWh Battery, 226-Mile Range. 2019. Available online: https://insideevs.com/nissan-reveals-leaf-e-plus-ces/ (Accessed: Nov 2023).
[26] Hyundai. All-New Hyundai NEXO—Technical Specifications. 2019. https://www.hyundai.news/eu/press-kits/all-new-hyundai-nexo-technical-specifications/ (Accessed: Nov 2023).
[27] U. Eberle, and R. Von Helmolt, "Sustainable transportation based on electric vehicle concepts: a brief overview", *Energy & Environmental Science*, vol. 3, no. 6, pp. 689–699, 2010.
[28] G. J. Offer, D. Howey, M. Contestabile, R. Clague, and N. P. Brandon, "Comparative analysis of battery electric, hydrogen fuel cell and hybrid vehicles in a future sustainable road transport system", *Energy Policy*, vol. 38, no. 1, pp. 24–29, 2010.
[29] X. Han, F. Li, T. Zhang, T. Zhang, and K. Song, "Economic energy management strategy design and simulation for a dual-stack fuel cell electric vehicle", *International Journal of Hydrogen Energy*, vol. 42, no. 16, pp. 11584–11595, 2017.
[30] H. El Fadil, F. Giri, J. M. Guerrero, and A. Tahri, "Modeling and nonlinear control of a fuel cell/supercapacitor hybrid energy storage system for

electric vehicles", *IEEE Transactions on Vehicular Technology*, vol. 63, no. 7, pp. 3011–3018, 2014.

[31] S. R. Fahim, and H. M. Hasanein, "Energy storage devices". in: A. F. Zobaa, S. H. E. Abdel Aleem (eds.), *Modernization of Electric Power Systems*, Springer, Cham, pp. 407–441, 2023.

[32] A. Millner, "Modeling lithium ion battery degradation in electric vehicles", *Conference on Innovative Technologies for an Efficient and Reliable Electricity Supply (CITRES)*, pp. 349–356, 2010.

[33] Alternative Energy. New Battery Technology Charges in Seconds. 2009. Available online: http://www.alternative-energy-news.info/new-battery-technology-charges-in-seconds/ (Accessed: Nov 2023).

[34] MIT Technology Review. A Battery for Electronics That Lasts Twice as Long. 2015. Available online: https://www.technologyreview.com/s/534626/a-battery-for-electronics-that-lasts-twice-as-long/ (Accessed: Nov 2023).

[35] J. Qian, W. A. Henderson, W. Xu, *et al.*, "High rate and stable cycling of lithium metal anode", *Nature Communications*, vol. 6, pp. 1–9, 2015.

[36] J. Kosivi, J. Gomez, R. Nelson, E. E. Kalu, and M. H. Weatherspoon, "Non-paste based composite cathode electrode for lithium air battery", in: *The Electrochemical Society Meeting Abstracts 225*, p. 206, 2014.

[37] Z. Zhao-Karger, and M. Fichtner, "Magnesium–sulfur battery: Its beginning and recent progress", *MRS Communications*, vol. 7, pp. 770–784, 2017.

[38] D. Gelman, B. Shvartsev, and Y. Ein-Eli, "Aluminum–air battery based on an ionic liquid electrolyte", *Journal of Materials Chemistry A*, vol. 47, pp. 20237–20242, 2014.

[39] Extremetech. Aluminium-Air Battery Can Power Electric Vehicles for 1000 Miles, Will Come to Production Cars in 2017. Available online: https://www.extremetech.com/extreme/151801-aluminium-air-battery-can-power-electric-vehicles-for-1000-mileswill-come-to-production-cars-in-2017 (Accessed: Nov 2023).

[40] Energy Storage Inter-Platform Group, State of the Art of Energy Storage Regulations and Technology. Available online: http://www.futured.es/wp-content/uploads/2016/06/GIA-Maqueta_eng.pdf (Accessed: Nov 2023).

[41] P. Adelhelm, P. Hartmann, C. L. Bender, M. Busche, C. Eufinger, and J. Janek, "From lithium to sodium: Cell chemistry of room temperature sodium–air and sodium–sulfur batteries", *Beilstein Journal of Nanotechnology*, vol. 6, pp. 1016–1055, 2015.

[42] Cleantechnica. Sodium-Air Batteries May Best Lithium-Air Batteries. 2013. Available online: https://cleantechnica.com/2013/0 3/20/sodium-air-batteries-may-best-lithium-air-batteries/ (Accessed: Nov 2023).

[43] Phys.org. Sodium-Air Battery Offers Rechargeable Advantages Compared to Li-Air Batteries. 2013. Available online: https://phys.org/news/2013-01-sodium-air-battery-rechargeable-advantages-li-air.html (Accessed: Nov 2023).

[44] O. A. Vargas-Ceballos, "*Estudio de Materiales Basados en Grafeno para su uso Como Ánodos en Baterías de Li-Ión", Ph.D. Thesis, Universidad de Córdoba*, Cordoba, Spain, 2013.

[45] F. García. La Española Graphenano Presenta una Batería que Dura 800 Kilómetros. 2016. Available online: http://www.elmundo.es/motor/2016/02/11/56bc7d6aca4741e31e8b461f.html (Accessed: Nov 2023).

[46] H. Kim, K. Y. Park, J. Hong, and K. Kang, "All-graphene-battery: Bridging the gap between supercapacitors and lithium ion batteries", *Science Reports*, vol. 4, pp. 1–8, 2014.

[47] T. Placke, A. Heckmann, R. Schmuch, P. Meister, K. Beltrop, and M. Winter, "Perspective on performance, cost, and technical challenges for practical dual-ion batteries", *Joule*, vol. 2, no. 12, pp. 2528–2550, 2018.

[48] X. Zhang, Y. Tang, F. Zhang, and C. S. Lee, "A novel aluminum–graphite dual-ion battery", *Advanced Energy Materials*, vol. 6, no. 11, pp. 1–6, 2016.

[49] X. Jiang, L. Luo, F. Zhong, *et al.*, "Electrolytes for dual-carbon batteries", *ChemElectroChem*, vol. 6, no. 10, pp. 2615–2629, 2019.

[50] J. Zhu, Y. Li, B. Yang, *et al.*, "A dual carbon-based potassium dual ion battery with robust comprehensive performance", *Small*, vol. 14, no. 31, pp. 1–6, 2018.

[51] S. Chen, Q. Kuang, and H. J. Fan, "Dual-carbon batteries: Materials and mechanism", *Small*, vol. 16, no. 40, pp. 1–13, 2020.

[52] A. A. Habib, M. K. Hasan, M. Mahmud, S. M. Motakabber, M. I. Ibrahimya, and S. Islam, "A review: Energy storage system and balancing circuits for electric vehicle application", *IET Power Electronics*, vol. 14, no. 1, pp. 1–13, 2021.

[53] D. Rajagopal, "Implications of the energy transition for government revenues, energy imports and employment: The case of electric vehicles in India", *Energy Policy*, vol. 175, pp. 1–10, 2023.

[54] A. Ahmadian, B. Mohammadi-Ivatloo, and A. Elkamel, "A review on plug-in electric vehicles: Introduction, current status, and load modeling techniques", *Journal of Modern Power Systems and Clean Energy*, vol. 8, no. 3, pp. 412–425, 2020.

[55] Kia e-Soul. Electric CUV, Kia Motors Worldwide, Retrieved 30 November 2020 from, https://www.kia.com/worldwide/vehicles/e-soul.do (Accessed: Nov 2023).

[56] M. A. Hannan, M. H. Lipu, A. Hussain, and A. Mohamed, "A review of lithium-ion battery state of charge estimation and management system in electric vehicle applications: Challenges and recommendations", *Renewable Sustainable Energy*, vol. 78, pp. 834–854, 2017.

[57] M. A. Hannan, M. M. Hoque, A. Mohamed, and A. Ayob, "Review of energy storage systems for electric vehicle applications: Issues and challenges", *Renewable Sustainable Energy*, vol. 69, pp. 771–789, 2017.

[58] L. Huang, Z. Zhang, Z. Wang, L. Zhang, X. Zhu, and D. D. Dorrell, "Thermal runaway behavior during overcharge for large-format lithium-ion batteries with different packaging patterns", *Journal of Energy Storage*, vol. 25, pp. 2–7, 2019.

[59] K. Shen, Q. Zhai, Y. Gu, *et al.*, "Life cycle assessment of lithium ion battery from water-based manufacturing for electric vehicles", *Resources, Conservation and Recycling*, vol. 198, 2023.

[60] V. Koval, Y. Sribna, O. Prokopenko, L. Filipishyna, V. Ivata, and I. W. Arsawan, "Management supply chains electric vehicle battery recycling", in: V. Koval, P. Olczak (eds.), *Circular Economy for Renewable Energy*, Springer, Cham, pp. 33–54, 2023.

[61] Z. Muhammad, Z. Anwar, B. Saleem, and J. Shahid, "Emerging cybersecurity and privacy threats to electric vehicles and their impact on human and environmental sustainability", *Energies*, vol. 16, no. 3, pp. 1–30, 2023.

[62] C. H. Basha, T. Mariprasath, M. Murali, and S. Rafikiran, "Simulative design and performance analysis of hybrid optimization technique for PEM fuel cell stack based EV application", *Materials Today: Proceedings*, vol. 52, no. 3, pp. 290–295, 2022.

[63] Electric Vehicles: A Small Window for Safety and Stability, a Large Arena for Research Forbes. Available online: https://www.forbes.com/sites/thebakersinstitute/2021/01/21/electric-vehicles-a-small-window-for-safety-and-stability-a-large-arena-for-research/?sh=134fd10b6f3 (Accessed: Nov 2023).

[64] F. G. Venegas, M. Petit, and Y. Perez, "Active integration of electric vehicles into distribution grids: Barriers and frameworks for flexibility services", *Renewable and Sustainable Energy Reviews*, vol. 145, pp. 1–57, 2021.

[65] J. Engelhardt, J. M. Zepter, T. Gabderakhmanova, and M. Marinelli, "Energy management of a multi-battery system for renewable-based high power EV charging", *eTransportation*, vol. 14, pp. 1–10, 2022.

[66] M. Moglia, J. Hopkins, and A. Bardoel, "Telework, hybrid work and the United Nation's Sustainable Development Goals: Towards policy coherence", *Sustainability*, vol. 13, no. 16, pp. 1–28, 2021.

[67] B. Anthony Jnr, S. Abbas Petersen, D. Ahlers, and J. Krogstie, "API deployment for big data management towards sustainable energy prosumption in smart cities-a layered architecture perspective", *International Journal of Sustainable Energy*, vol. 39, no. 3, pp. 263–289, 2020.

[68] W. Sun, F. Neumann, and G. P. Harrison, "Robust scheduling of electric vehicle charging in LV distribution networks under uncertainty", *IEEE Transactions on Industry Applications*, vol. 56, no. 5, pp. 5785–5795, 2020.

[69] R. Wolbertus, S. Jansen, and M. Kroesen, "Stakeholders' perspectives on future electric vehicle charging infrastructure developments", *Futures*, vol. 123, pp. 1–22, 2020.

[70] S. Mohanty, S. Panda, S. M. Parida, *et al.*, "Demand side management of electric vehicles in smart grids: A survey on strategies, challenges, modeling, and optimization", *Energy Reports*, vol. 8, pp. 12466–12490, 2022.

[71] K. Prakash, M. Ali, M. N. Siddique, *et al.*, "Bi-level planning and scheduling of electric vehicle charging stations for peak shaving and congestion management in low voltage distribution networks", *Computers and Electrical Engineering*, vol. 102, 2022.

[72] D. La Gamba, M. Pirra, F. Deflorio, L. Montesano, A. Carboni, and M. Arnone, "Discrete Choice Experiments to identify user preference for electric mobility", *2022 IEEE 46th Annual Computers, Software, and Applications Conference (COMPSAC)*, pp. 1696–1701, 2022.

[73] F. Ahmad, M. S. Alam, S. M. Shariff, and M. Krishnamurthy, "A cost-efficient approach to EV charging station integrated community microgrid: A case study of Indian power market", *IEEE Transactions on Transportation Electrification*, vol. 5, no. 1, pp. 200–214, 2019.

[74] S. Borozan, S. Giannelos, and G. Strbac, "Strategic network expansion planning with electric vehicle smart charging concepts as investment options", *Advances in Applied Energy*, vol. 5, pp. 1–19, 2022.

[75] S. Mustafa, W. Zhang, and R. Li, "Does environmental awareness play a role in EV adoption? A value-based adoption model analysis with SEM-ANN approach", *IEEE/WIC/ACM International Conference on Web Intelligence and Intelligent Agent Technology*, pp. 433–440, 2021.

[76] H. Zhao, X. Yan, and H. Ren, "Quantifying flexibility of residential electric vehicle charging loads using non-intrusive load extracting algorithm in demand response", *Sustainable Cities and Society*, vol. 50, pp. 1–12, 2019.

[77] B. Shirizadeh, and P. Quirion, "The importance of renewable gas in achieving carbon-neutrality: Insights from an energy system optimization model", *Energy*, vol. 255, pp. 1–49, 2022.

[78] M. T. Kahlen,W. Ketter, and J. van Dalen, "Electric vehicle virtual power plant dilemma: Grid balancing versus customer mobility", *Production and Operations Management*, vol. 27, no. 11, pp. 2054–2070, 2018.

[79] Z. Huang, Z. Guo, P. Ma, *et al.*, "Economic-environmental scheduling of microgrid considering V2G-enabled electric vehicles integration", *Sustainable Energy, Grids and Networks*, vol. 32, 2022.

[80] P. Jamborsalamati, M. J. Hossain, S. Taghizadeh, G. Konstantinou, M. Manbachi, and P. Dehghanian, "Enhancing power grid resilience through an IEC61850-based EV-assisted load restoration", *IEEE Transactions on Industrial Informatics*, vol. 16, no. 3, pp. 1799–1810, 2019.

[81] S. Sharma, and P. Jain, "Integrated TOU price-based demand response and dynamic grid-to-vehicle charge scheduling of electric vehicle aggregator to support grid stability", *International Transactions on Electrical Energy Systems*, vol. 30, no. 1, pp. 1–22, 2020.

[82] S. Shaheen, L. Cano, and M. Camel, "Exploring electric vehicle carsharing as a mobility option for older adults: A case study of a senior adult community in the San Francisco Bay Area", *International Journal of Sustainable Transportation*, vol. 10, no. 5, pp. 406–417, 2016.

[83] Y. Zhou, S. Cao, J. L. Hensen, and P. D. Lund, "Energy integration and interaction between buildings and vehicles: A state-of-the-art review", *Renewable and Sustainable Energy Reviews*, vol. 114, pp. 1–25, 2019.

[84] R. Alvaro-Hermana, J. Fraile-Ardanuy, P. J. Zufiria, L. Knapen, and D. Janssens, "Peer to peer energy trading with electric vehicles", *IEEE Intelligent Transportation Systems Magazine*, vol. 8, no. 3, pp. 33–44, 2016.

[85] A. M. Farid, A. Viswanath, R. Al-Junaibi, D. Allan, and T. J. Van der Wardt, "Electric vehicle integration into road transportation, intelligent transportation, and electric power systems: an Abu Dhabi case study", *Smart Cities*, vol. 4, no. 3, pp. 1039–1057, 2021.

[86] J. G. Walters, S. Marsh, and L. Rodrigues, "Planning perspectives on rural connected, autonomous and electric vehicle implementation", *Sustainability*, vol. 14, no. 3, pp. 1–20, 2022.

[87] J. M. Clairand, P. Guerra-Terán, X. Serrano-Guerrero, M. González-Rodríguez, and G. Escrivá-Escrivá, "Electric vehicles for public transportation in power systems: A review of methodologies", *Energies*, vol. 12, no. 16, pp. 1–22, 2019.

[88] E. S. Rigas, S. D. Ramchurn, and N. Bassiliades, "Managing electric vehicles in the smart grid using artificial intelligence: A survey", *IEEE Transactions on Intelligent Transportation Systems*, vol. 16, no. 4, pp. 1619–1635, 2014.

[89] S. S. Shuvo, and Y. Yilmaz, "Predictive maintenance for increasing EV charging load in distribution power system", *2020 IEEE International Conference on Communications, Control, and Computing Technologies for Smart Grids (SmartGridComm)*, pp. 1–6, 2020.

[90] M. Shibl, L. Ismail, and A. Massoud, "Electric vehicles charging management using machine learning considering fast charging and vehicle-to-grid operation", *Energies*, vol. 14, no. 19, 2021.

[91] A. Petrucci, G. Barone, A. Buonomano, and A. Athienitis, "Modelling of a multi-stage energy management control routine for energy demand forecasting, flexibility, and optimization of smart communities using a Recurrent Neural Network", *Energy Conversion and Management*, vol. 268, 2022.

[92] D. Jozwiak, J. R. Pillai, P. Ponnaganti, B. Bak-Jensen, and J. Jantzen, "Optimal charging and discharging strategies for electric cars in PV-BESS-based marina energy systems", *Electronics*, vol. 12, no. 4, pp. 1–21, 2023.

[93] S. Tirunagari, M. Gu, and L. Meegahapola, "Reaping the benefits of smart electric vehicle charging and vehicle-to-grid technologies: Regulatory, policy and technical aspects", *IEEE Access*, vol. 10, 2022.

[94] M. A. Hasan, R. Chapman, and D. J. Frame, "Acceptability of transport emissions reduction policies: A multi-criteria analysis", *Renewable and Sustainable Energy Reviews*, vol. 133, pp. 1–12, 2020.

[95] S. F. da Silva, J. J. Eckert, F. L. Silva, L. C. Silva, and F. G. Dedini, "Multi-objective optimization design and control of plug-in hybrid electric vehicle powertrain for minimization of energy consumption, exhaust emissions and battery degradation", *Energy Conversion and Management*, vol. 234, pp. 1–20, 2021.

[96] H. A. Ayad, L. Ala'Al-Kilani, R. Arshad, M. A. Al-Obadi, H. T. Hussein, and M. Kucukvar, "Developing an interactive data visualization platform to present the adaption of electrical vehicles in Washington, California and New York", *2020 IEEE 7th International Conference on Industrial Engineering and Applications (ICIEA)*, pp. 731–737, 2020.

[97] A. Dargahi, S. Ploix, A. Soroudi, and F. Wurtz, "Optimal household energy management using V2H flexibilities", *COMPEL: The International Journal for Computation and Mathematics in Electrical and Electronic Engineering*, vol. 33, no. 3, pp. 777–792, 2014.

[98] Z. Yi, and J. Smart, "A framework for integrated dispatching and charging management of an autonomous electric vehicle ride-hailing fleet", *Transportation Research Part D: Transport and Environment*, vol. 95, pp. 1–52, 2021.

[99] S. Micari, A. Polimeni, G. Napoli, L. Andaloro, and V. Antonucci, "Electric vehicle charging infrastructure planning in a road network", *Renewable and Sustainable Energy Reviews*, vol. 80, pp. 98–108, 2017.

[100] S. Acharya, Y. Dvorkin, H. Pandžić, and R. Karri, "Cybersecurity of smart electric vehicle charging: A power grid perspective", *IEEE Access*, vol. 8, pp. 214434–214453. 2020.

[101] S. Zhang, and J. Q. James, "Electric vehicle dynamic wireless charging system: Optimal placement and vehicle-to-grid scheduling", *IEEE Internet of Things Journal*, vol. 9, no. 8, pp. 6047–6057, 2021.

[102] T. Hu T, K. Liu, and H. Ma, "Probabilistic electric vehicle charging demand forecast based on deep learning and machine theory of mind", *2021 IEEE Transportation Electrification Conference & Expo (ITEC)*, pp. 795–799, 2021.

[103] A. Viziteu, D. Furtună, A. Robu, *et al.*, "Smart scheduling of electric vehicles based on reinforcement learning", *Sensors*, vol. 22, no. 10, pp. 1–14, 2022.

[104] Y. Shen, W. Fang, F. Ye, and M. Kadoch, "EV charging behavior analysis using hybrid intelligence for 5G smart grid", *Electronics*, vol. 9, no. 1, pp. 1–17, 2020.

[105] W. Yin, Z. Ming, and T. Wen, "Scheduling strategy of electric vehicle charging considering different requirements of grid and users", *Energy*, vol. 232, pp. 1–9, 2021.

[106] J. Hu, A. Saleem, S. You, *et al.*, "A multi-agent system for distribution grid congestion management with electric vehicles", *Engineering Applications of Artificial Intelligence*, vol. 38, pp. 45–58, 2015.

[107] Z. Ullah, A. U. Rehman, S. Wang, *et al.*, "IoT-based monitoring and control of substations and smart grids with renewables and electric vehicles integration", *Energy*, vol. 282, 2023.

[108] T. Jiang, G. Deconinck, J. Wu, *et al.*, "Guest editorial: Special issue on integrated local energy systems", *International Journal of Electrical Power & Energy Systems*, vol. 148, 2023.

[109] H. Liu, X. Zhang, X. Shen, H. Sun, and M. Shahidehpour, "A hybrid federated learning framework with dynamic task allocation for multi-party distributed load prediction", *IEEE Transactions on Smart Grid*, vol. 14, no. 13, pp. 2460–2472, 2022.

Chapter 5

Optimization-based parameter estimation for PEM fuel cells in transportation applications

Salma G. Saad Eldin[1], Essam El-Din Abou El-Zahab[1], Shady H.E. Abdel Aleem[2] and Mohamed M. Refaat[3]

This chapter focuses on the parameter estimation of proton exchange membrane fuel cells (PEMFCs) in the context of transportation applications. With the increasing demand for clean and sustainable energy sources, PEMFCs have emerged as a promising technology due to their high efficiency, low operating temperature, and environmental friendliness. Accurate parameter estimation plays a crucial role in achieving reliable simulation and performance prediction of PEMFC systems. Various optimization-based methods have been employed to estimate the parameters of PEMFCs, considering the non-linear nature of the equations and characteristic curves involved. In this chapter, we explore the application of metaheuristic optimization algorithms. Besides, this chapter provides insights into the optimization-based techniques employed in the transportation sector to enhance the efficiency and reliability of PEMFCs, contributing to the development of sustainable power generation in transportation applications.

5.1 Introduction

As a result of technological and economic progress, environmental changes, and social and political actions, the use of renewable energy sources (RES) to provide clean and sustainable energy is fast expanding [1]. At the same time, traditional fossil fuel-based technologies confront several problems, including price changes, severe environmental consequences from greenhouse gas emissions and global warming, and global fuel resource depletion [2]. In this regard, implementing rigorous environmental policies and the enforcement of strict regulations on carbon emissions have sparked a growing interest in developing green power production technologies [3]. In recent years, there has been an increased focus on fuel cell (FC)

[1]Electrical Power and Machines Engineering, Faculty of Engineering, Cairo University, Egypt
[2]Department of Electrical Engineering, Institute of Aviation Engineering and Technology, Egypt
[3]Photovoltaic Cells Department, Electronics Research Institute, Egypt

systems due to factors such as their high efficiency, high power density, low operating temperature, little to no noise, good dynamic response, and low environmental impact. Compared to other forms of RESs, FCs are considered dependable and long-lasting electrochemical devices that consistently convert hydrogen fuel into electricity. FC deployment in electrical engineering contributes to sustainable power generation, improves energy resiliency, and reduces reliance on conventional fossil fuel-based electricity generation methods [4–6].

FCs can be classified into different types based on the form of electrolyte used [7]. The main types of FCs based on electrolytes are [8]:

1. Proton exchange membrane fuel cells (PEMFCs) [9,10],
2. Solid oxide FCs [11,12],
3. Molten carbonate FCs [13,14],
4. Phosphoric acid FCs [15,16],
5. Alkaline FCs [17,18], and
6. Direct methanol FCs [17,18].

Each FC type has its merits, disadvantages, and specific applications. The FC type selected is determined by parameters such as the intended operating temperature, fuel availability, power output requirements, and system constraints [7]. PEMFCs offer several advantages that make them attractive for various applications [19].

Figure 5.1 illustrates the primary keywords used in publications centered on FCs and their distribution. This visual representation aids in understanding the primary research emphasis of the analyzed papers, facilitating comprehension of the areas of expertise. Furthermore, Figure 5.1 offers guidance for prospective

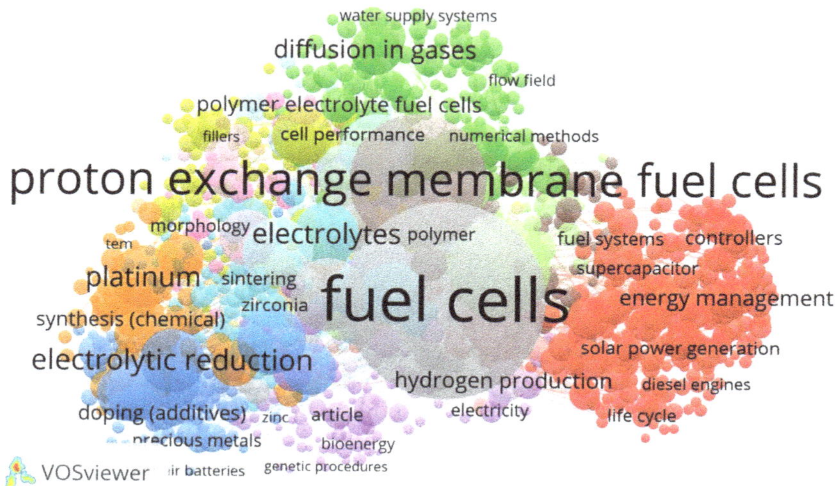

Figure 5.1 Co-occurrence analysis map of the articles focusing on PEM in FC

research endeavors. The dimensions of the spheres are indicative of the relative frequency of terms. To clarify, it may be inferred that more giant clouds and patterns indicate an increased frequency of occurrence. Furthermore, Figure 5.2 depicts the distribution of publications in specific countries. China and the United States were found to have the largest number of publications in the specific field of research under consideration.

PEMFCs have high energy efficiency, typically ranging from 40% to 60%. This means they can transform a considerable percentage of the chemical energy in the fuel into electrical energy, leading to lower fuel consumption and greenhouse gas emissions [20,21]. PEMFCs can be rapidly started and shut down, allowing on-demand power generation. This feature is precious in applications requiring intermittent or frequent power cycling such as portable devices and automotive applications. PEMFCs operate at relatively low temperatures, typically below 100 degrees Celsius. This enables faster warm-up times, reduces the need for extensive insulation, and simplifies thermal management. The lower operating temperature also promotes longer system durability and extends the life expectancy of the FC stack. PEMFCs are lightweight and compact, making them suitable for portable and mobile applications. They have a high power-to-weight ratio, allowing for efficient power generation without adding significant weight or bulk to the overall system.

PEMFCs can utilize a variety of fuels, with pure hydrogen being the most common fuel source [22,23]. However, they can also operate on hydrogen-rich gases such as reformat gas from hydrocarbon fuels like natural gas or methanol. This fuel flexibility provides options for fuel availability and enables the use of renewable hydrogen sources. PEMFCs produce virtually zero emissions at the point of use. Water vapor and

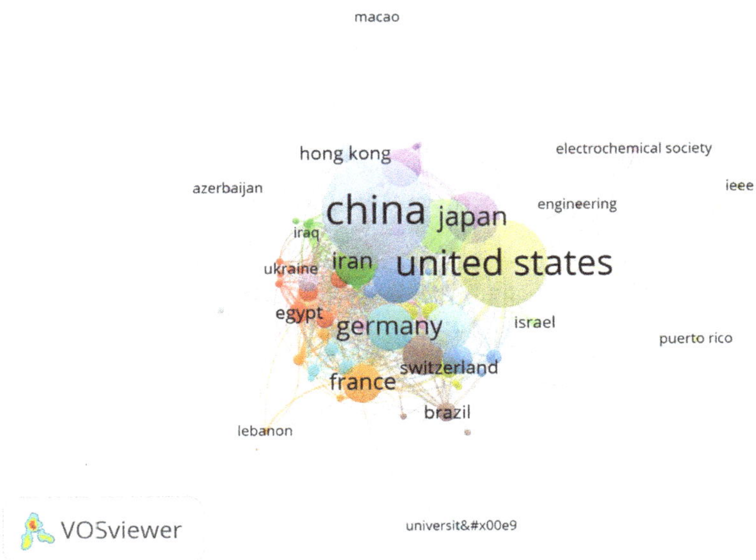

Figure 5.2 PEM's article distributions concerning countries

heat are the only byproducts of the electrochemical reaction in a PEMFC. This makes them environmentally friendly and helps to improve air quality, particularly when they use renewable hydrogen sources for power. PEMFCs can be flexibly scaled up or down to fulfill a variety of power needs. They can be found in tiny, portable devices and bigger applications like residential, commercial, and automotive systems. Given its scalability, PEMFCs are adaptable and flexible to a variety of power generation applications. PEMFC technology has seen significant advancements in terms of durability and reliability. Improved catalyst materials, membrane materials, and cell designs have enhanced the longevity and performance of PEMFCs, making them more suitable for long-term operation and commercial viability [24].

Accurate parameter estimation of the FC model is regarded as a crucial phase in achieving a reliable simulation of the FC system's features [25,26]. A considerable quantity of design parameters remain undetermined, necessitating exact calculations. Researchers encounter a significant obstacle when examining a mathematical model of PEMFCs to study certain functions and their corresponding polarization curves. Statistical information in a PEMFC datasheet is typically insufficient for identifying the optimal parameter selections [27,28].

Nevertheless, without precise parameters, significant disparities arise between the data derived from the model and the information provided in the manufacturer's datasheet. Parameter estimation in PEMFCs involves determining the values of various parameters that characterize the behavior and performance of the FC. These parameters include the cell voltage, current density, electrochemical reaction rates, transport properties, and other factors that affect the FC's efficiency and durability [29].

Understanding the precise PEMFC cell model is necessary to examine the operating principle, attributes, and potential for operation at the point of maximum power [30]. The literature distinguishes between static and dynamic PEMFC cell models [31,32]. While dynamic models are essential for the functioning of devices in complex systems with real and frequent transient processes, static models describe the behavior of PEMFC cells in a steady state. The Mann *et al.* [33] model is the most used PEMFC cell model for static models. The next section will describe how the sum of many distinct voltage drops represents the voltage at the ends of the PEMFC cell in this model. As a result, the Mann *et al.* model clearly demonstrates the relationship between the PEMFC cell's voltage and current. The fact that this relationship is a highly non-linear function should be emphasized [32].

There are numerous ways to estimate parameters in PEM FCs, and the specific aims of the investigation and the available data determine the chosen strategy. For illustration, experimental data fitting, mathematical modeling, system identification, sensitivity analysis, and optimization methods are commonly employed in the literature to extract PEM parameters. The optimization-based methods are the subject of this chapter.

Parameter estimation in PEM FCs is complex and challenging, and selecting a suitable optimization algorithm is crucial [25,26]. Due to the non-linear structure of the relevant equations and characteristic curves, standard approaches face challenges in solving them. Consequently, employing analytical methods to address the problem becomes burdensome. As a result, metaheuristic techniques have been commonly utilized to estimate FC parameters [34]. Several recent metaheuristics have been applied in the literature to solve the PEMFC problem. For instance, the slime mould algorithm was applied by Gupta *et al.* [35] to estimate the PEMFC model. The obtained results demonstrated superior performance compared to the evaluated algorithms. The authors of [36] introduced a novel optimization technique, the Hybrid Grey Wolf Optimizer. The algorithm under consideration is a novel metaheuristic approach that draws inspiration from the collective behavior exhibited by gray wolf packs. The comprehensive analysis of several benchmarks demonstrated that the Hybrid Grey Wolf Optimizer approach exhibits high efficiency across all evaluated criteria, including convergence and accuracy.

Additionally, the use of the Hybrid Grey Wolf Optimizer was employed to estimate the model of PEMFC. The black widow optimization (BWO) algorithm was developed by Singla *et al.* [37]. It was applied to extract the parameters of models for PEMFCs across varying operating temperatures. The results of BWO were contrasted to those obtained from five other algorithms, namely Particle Swarm Optimization (PSO), Multi-Verse Optimizer (MVO), Sine Cosine Method (SCA), Whale Optimization Algorithm (WOA), and Grey Wolf Optimization (GWO).

Xu *et al.* [38] suggested a straightforward two-stage approach for accurately predicting the unidentified model parameters of PEMFCs. The technique combines the JAYA and Nelder-Mead simplex algorithms, improving estimation performance. The efficacy of the suggested technique was assessed using estimation experiments including seven and nine parameters with unknown values. The proposed strategy demonstrated superior accuracy, convergence speed, and stability performance compared to the basic JAYA algorithm and four other recently published meta-heuristic algorithms (Grasshopper Optimization Algorithm, Salp Swarm Optimizer, and MVO). Consequently, it exhibits great potential as an effective approach for parameter estimation of PEMFCs. The study [39] presented an optimal parameter estimation for PEMFCs using the improved Red Fox optimizer. The performance evaluation of the proposed algorithm demonstrated favorable results compared to its preceding state-of-the-art techniques.

The authors in [40] introduced an improved approach for estimating the parameters of unknown variables in the model of PEMFECs. The aim was to minimize the absolute error between the estimated data obtained using the proposed model and the actual data. The suggested methodology employed an enhanced version of the Archimedes Optimization Algorithm for this objective. The final results proved that the proposed method yields the most favorable outcome compared to other methods. A novel approach for developing an ideal dynamic model for PEMFCs was suggested by Duan *et al.* [41]. The primary focus of this study pertained to the best selection of uncertain parameters within the dynamic model of a PEMFC. The primary aim was to reduce the sum of the squared error between the voltage predicted by the model and the voltage obtained from experimental data. The minimizing strategy relied on utilizing a modified iteration of the deer hunting optimization algorithm. The results were compared with experimental data and other contemporary methodologies. In Zaki *et al.* [42], the Marine Predators Algorithm (MPA) and Political Optimizer (PO) were tested to assess the unidentified parameters of PEMFC models. The optimization techniques have been tested on three commercial PEMFC stacks. The results showed that the PEMFC model utilizing the MPA technique outperformed all other algorithms examined in regard to both solution accuracy and convergence speed. Abaza *et al.* [43] introduced a unique swarm-based approach known as the Coyote Optimization Approach (COA) to determine the best parameter for both PEMFCs. Two practical PEMFCs were used as models to assess the effectiveness of the proposed algorithm across various operating situations. The efficacy of the COA was proven by conducting a comparative analysis with four optimizers in identical settings.

Houssein *et al.* [44] introduced a modified version of the artificial electric field algorithm (AEFA) to determine the optimal parameters for PEMFCs effectively. The advantage of the proposed algorithm was demonstrated by utilizing two distinct types of PEMFCs. Based on the primary findings, the efficacy of the developed optimizer in evaluating the optimal parameters for PEMFC was proved compared to alternative optimization methods. Fathy *et al.* [45] introduced a novel approach, termed the LSHADE-EpSin optimization algorithm, which combines ensemble sinusoidal parameter adaptation with L-SHADE. The suggested methodology aimed to address the parameter extraction problem in PEMFC. The findings demonstrated the efficacy of the suggested methodology utilizing the LSHADE-EpSin algorithm in accurately calculating the optimal parameters of the PEMFC across various operational scenarios compared to alternative optimization methods. Although metaheuristic techniques are powerful optimization methods for solving PEM problems, they also come with challenges like convergence, scalability, parameter tuning, interpretability, computational resources, uncertainty, and problem formulation. Metaheuristic techniques can get stuck in local optima, which may prevent them from finding the global optimum. This can happen when the search space is large and complex, and the algorithm cannot explore all the possible solutions. Artificially intelligent (AI) systems are used to extract PEMFC parameters [46]. Recent research has shown the effectiveness of AI in solving the PEMFC problem. Using the support vector machine artificial intelligence approach, Peng *et al.* [41] developed an ideal power density model for PEMFCs. The results obtained confirmed the efficiency of the proposed approach. Artificial neural network (ANN) models, alongside process simulators, are well-recognized methodologies for estimating or optimizing parameters in gasification processes, FCs, and other integrated systems. Zhang *et al.* [47] created the multiple learning neural network algorithm (MLNNA), which is an improved version of the neural network algorithm (NNA). The goal was to find the unknown parameters of PEMFC models accurately. The results of the experiments show that MLNNA is better at estimating parameters for PEMFC models. This was shown by comparing it to ten other robust optimization techniques. In Kartal *et al.* [48], the ANN was used to estimate PEMFC's parameters, and the findings demonstrated a high level of accuracy of ANN in predicting the output characteristics of PEMFCs. It should be mentioned that machine learning and multi-objective and hybrid optimization advances are driving the future of PEMFC optimization, promising more sustainable and efficient energy conversion. Figure 5.3 shows a brief representation of the future trends in PEMFC studies.

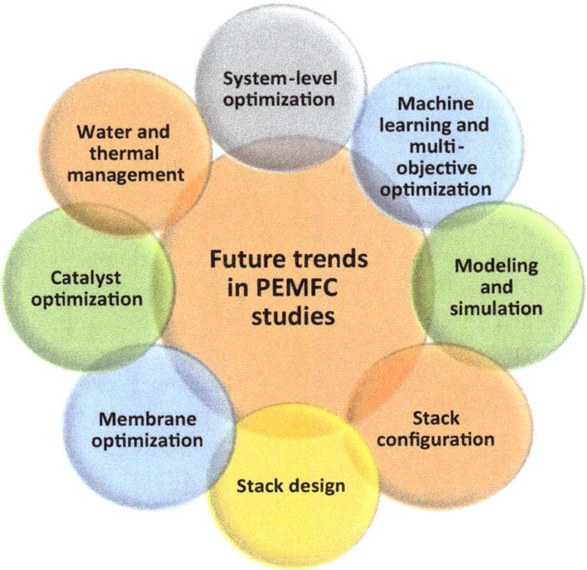

Figure 5.3 Future trends in PEMFC studies

From an optimization standpoint, anticipated future trends in PEMFC studies focus on enhancing PEMFC performance, efficiency, and cost-effectiveness:

1. System-level optimization: Researchers seek to maximize efficiency and performance by optimizing the overall system design and operation of PEMFCs. This entails minimizing losses and improving overall system efficiency by optimizing the size and design of various components such as the stack, plant balance, and cooling system.
2. Control strategies: Advanced control methods must be developed to optimize PEMFC operation. Future research could concentrate on developing intelligent control algorithms that adapt to dynamic operating situations, manage transitory behavior, and optimize FC performance in real time.
3. Adequate water and thermal management are crucial for optimizing PEMFC performance. Future research will create sophisticated ways to control humidity levels, prevent water floods or dehydration, and improve heat transport within the FC system. Optimizing the design of flow channels, membranes, and thermal management components is part of this.
4. Catalyst optimization: Catalysts are essential in the electrochemical reactions within PEMFCs. Future research will try to improve reaction kinetics and reduce the amount of expensive catalyst materials required by optimizing catalyst loading, particle size, and distribution. This optimization has the potential to increase efficiency and cost-effectiveness.
5. PEMFC membranes directly influence the proton conductivity and durability of the FC. In the future, membrane parameters such as thickness, ionomer

content, and porosity will be optimized to improve proton transfer and reduce membrane resistance. Furthermore, developments in membrane manufacturing processes may allow for more cost-effective production.

6. Stack design and configuration: By optimizing the design and configuration of FC stacks, power density, efficiency, and durability can be increased. Future research will most likely optimize gas distribution, lower contact resistance, and boost overall stack performance using novel stack topologies and materials.
7. Modeling and simulation: In FC research, computational optimization approaches are critical. In the future, improved models and simulation tools will be developed to correctly anticipate FC performance, degradation processes, and system behavior. To improve performance, these models can optimize numerous design and operating characteristics.
8. Multi-objective optimization: Considering many competing objectives is critical in FC optimization. Future research may use multi-objective optimization approaches and machine-learning-based algorithms to optimize efficiency, power density, durability, and cost simultaneously, allowing researchers to examine trade-offs and uncover optimal design and operating points.

These optimization trends seek to improve PEMFC performance, efficiency, durability, and cost-effectiveness, making them more suited for widespread commercial applications in various industries, including transportation and stationary power production. Remember that correct parameter extraction is required for accurate PEMFC modeling and simulation, which aids in their optimization and practical implementation.

In the context of optimizing PEMFCs, the objective functions typically revolve around improving specific performance metrics. Some standard objective functions are to maximize power density (aims to achieve the highest power output per unit area of the PEMFC), maximize efficiency (to increase the overall energy conversion efficiency of the PEMFC system), minimize voltage losses (to reduce voltage losses due to factors like ohmic, activation, and concentration losses), minimize cost per unit power (to achieve the best balance between cost and performance, ensuring cost-effective PEMFC systems), optimize mass transport (to minimize reactant concentration gradients and to ensure efficient transport of hydrogen and oxygen to the electrode surfaces), maintain stable operation (for ensuring stable and reliable performance over time, even under varying operating conditions), maximize durability and longevity (to extend the lifespan of PEMFC components and reducing degradation), balance water management (to prevent flooding or drying of the membrane, which can affect performance), and finally adaptation to specific applications (for tailoring the optimization objectives based on the intended application, whether it is for automotive, stationary power, or portable devices). These objective functions serve as the guiding criteria for the optimization process, helping to fine-tune the design and operating parameters of PEMFC systems for specific performance goals. Referring to metrics used to assess the quality of parameter extraction in the context of PEMFCs, common metrics that may be used:

- Root mean square error (*RMSE*): The *RMSE* measures the average magnitude of the errors between observed and predicted values, with higher errors receiving more weight as shown in (1) [49]. x_i and $\widehat{x}_i$ are the measured raw data and predicted data of the PEMFC voltage values, respectively. n is the number of points of measured voltages. However, *RMSE* is susceptible to outliers or large errors and penalizes greater errors severely. It is often utilized in regression projects and is especially beneficial when high errors are a concern.

$$\text{RMSE} = \sqrt{\frac{1}{n}\sum_{i=1}^{n}(x_i - \widehat{x}_i)^2} \tag{5.1}$$

- Mean absolute error (*MAE*): *MAE* calculates the average absolute difference between actual and expected values, offering a simple estimate of error magnitude as explained in (2) [50]. When compared to *RMSE*, it is less sensitive to extremes. Treats all errors identically, regardless of size. The average prediction accuracy is indicated.

$$\text{MAE} = \frac{1}{n}\sum_{i=1}^{n}|x_i - \widehat{x}_i| \tag{5.2}$$

- Mean absolute percentage error (MAPE): MAPE is a metric that quantifies the precision of a forecasting technique. It is the average of the absolute percentage errors of each entry in a dataset and is used to calculate the accuracy of the forecasted quantities relative to the actual quantities as illustrated by (3). MAPE is frequently effective for analyzing large datasets and requires non-zero values in the dataset.

$$\text{MAPE} = \frac{1}{n}\sum_{i=1}^{n}\left|\frac{x_i - \widehat{x}_i}{x_i}\right| \tag{5.3}$$

- Sum of squared errors (*SSE*): It is the sum of the squared deviations between observed and anticipated values, and it can be calculated by (4) [51]. It calculates the overall size of mistakes without considering the number of data points. It does not calculate the average error per data point directly. Because greater errors add disproportionately to the total, it is vulnerable to outliers. *SSE* is frequently used in optimization algorithms to identify the parameter values with the lowest error. Furthermore, SSE focuses on greater errors and is utilized in optimization, but *MAE* provides a more interpretable measure of average error magnitude and is less sensitive to outliers. The context and goals of the modeling or parameter extraction procedure influence the decision between *SSE* and *MAE*.

$$\text{SSE} = \sum_{i=1}^{n}(x_i - \widehat{x}_i)^2 \tag{5.4}$$

- Coefficient of determination (*R*-squared): *R*-squared reflects the proportion of the variance in the dependent variable that is predictable from the independent variables as described in (5) [52]. Gives a measure of the goodness of fit. It ranges from 0 to 1, with higher numbers indicating a better fit. This metric measures how well the model captures data variability.

$$R - \text{squared} = 1 - \sum_{i=1}^{n} (x_i - \widehat{x}_i)^2 / \sum_{i=1}^{n} (x_i - \overline{x}_i)^2 \tag{5.5}$$

- Normalized root mean square error (*NRMSE*): *NRMSE* normalizes the error with the data range as presented by (5.6) [53]. $NRMSE_{\max}$ and $NRMSE_{\min}$ are the maximum and minimum values of the RMSE, respectively. This is useful for comparing models across datasets. This is especially useful when comparing models applied to different datasets with varied sizes.

$$\text{NRMSE} = (NRMSE - NRMSE_{\min})/(NRMSE_{\max} - NRMSE_{\min}) \tag{5.6}$$

These general metrics are critical for determining the quality and dependability of parameter extraction methods used in PEMFC modeling and simulation. They also serve various functions and are selected based on the specific context and goals of the modeling or parameter extraction procedure. The metric chosen determines which features of the model's performance are most important for the specific application.

From the literature discussion presented above, it is clear that there is a necessity for alternative energy sources to meet the world's expanding energy demand, which has piqued academics' interest in FCs. The utilization of PEMFC relies heavily on its *V-I* properties, and the PEMFC parameters heavily influence its characteristics. To generalize, there is always an error when comparing FC parameter values acquired through optimization approaches to experimental ones. Figure 5.4 depicts the parameter estimate procedure. The primary goal of researchers in diverse literature is to reduce this error.

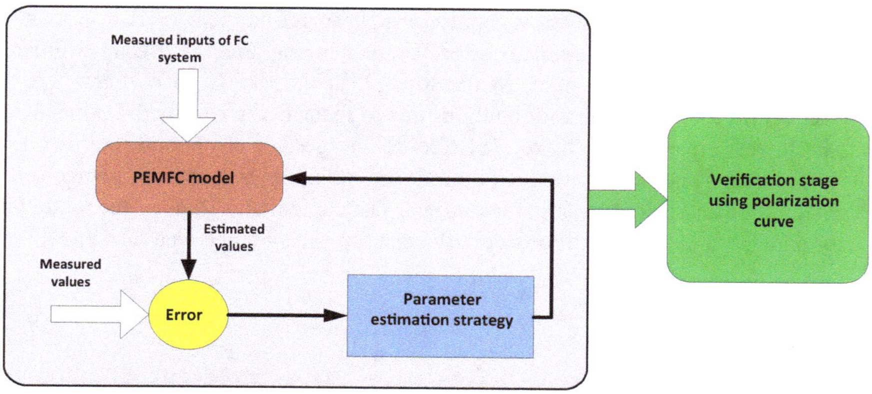

Figure 5.4 The procedure of PEMFC parameter estimation

References

[1] Ali ZM, Calasan M, Aleem SHEA, Jurado F, and Gandoman FH. Applications of energy storage systems in enhancing energy management and access in microgrids: A review. *Energies* 2023;16: 5930. https://doi.org/10.3390/en16165930.

[2] Eshra NM, Zobaa AF, and Abdel Aleem SHE. Assessment of mini and micro hydropower potential in Egypt: Multi-criteria analysis. *Energy Reports* 2021; 7:81–94. https://doi.org/10.1016/j.egyr.2020.11.165.

[3] Felseghi R-A, Carcadea E, Raboaca MS, Trufin CN, and Filote C. Hydrogen fuel cell technology for the sustainable future of stationary applications. *Energies* 2019;12:4593.

[4] Sorlei I-S, Bizon N, Thounthong P., *et al.* Fuel cell electric vehicles—A brief review of current topologies and energy management strategies. *Energies* 2021;14:252.

[5] Gong X, Dong F, Mohamed MA, Abdalla OM, and Ali ZM. A secured energy management architecture for smart hybrid microgrids considering PEM-fuel cell and electric vehicles. *IEEE Access* 2020;8:47807–23. https://doi.org/10.1109/ACCESS.2020.2978789.

[6] İnci M, Büyük M, Demir MH, and İlbey G. A review and research on fuel cell electric vehicles: Topologies, power electronic converters, energy management methods, technical challenges, marketing and future aspects. *Renew Sustain Energy Rev* 2021;137:110648.

[7] Abdelkareem MA, Elsaid K, Wilberforce T, Kamil M, Sayed ET, and Olabi A. Environmental aspects of fuel cells: A review. *Sci Total Environ* 2021; 752:141803.

[8] Ramezanizadeh M, Alhuyi Nazari M, Hossein Ahmadi M, and Chen L. A review on the approaches applied for cooling fuel cells. *Int J Heat Mass Transf* 2019; 139:517–25. https://doi.org/10.1016/j.ijheatmasstransfer.2019.05.032.

[9] Chen M, Zhao C, Sun F, Fan J, Li H, and Wang H. Research progress of catalyst layer and interlayer interface structures in membrane electrode assembly (MEA) for proton exchange membrane fuel cell (PEMFC) system. *ETransportation* 2020;5:100075.

[10] Okonkwo PC, Ige OO, Uzoma PC, Emori W, Benamor A, and Abdullah AM. Platinum degradation mechanisms in proton exchange membrane fuel cell (PEMFC) system: A review. *Int J Hydrogen Energy* 2021;46:15850–65.

[11] Singh M, Zappa D, and Comini E. Solid oxide fuel cell: Decade of progress, future perspectives and challenges. *Int J Hydrogen Energy* 2021;46:27643–74.

[12] Xu Q, Guo Z, Xia L., *et al.* A comprehensive review of solid oxide fuel cells operating on various promising alternative fuels. *Energy Convers Manag* 2022;253:115175.

[13] Mehrpooya M, Raeesi M, Pourfayaz F, and Delpisheh M. Investigation of a hybrid solar thermochemical water-splitting hydrogen production cycle and coal-fueled molten carbonate fuel cell power plant. *Sustain Energy Technol Assessments* 2021;47:101458.

[14] Wejrzanowski T, Cwieka K, Skibinski J., *et al.* Microstructure driven design of porous electrodes for molten carbonate fuel cell application: Recent progress. *Int J Hydrogen Energy* 2020;45:25719–32.

[15] Cheng S, Zhao G, Gao M, Shi Y, Huang M, and Marefati M. A new hybrid solar photovoltaic/phosphoric acid fuel cell and energy storage system; energy and exergy performance. *Int J Hydrogen Energy* 2021; 46:8048–66.

[16] Sun Q, Lin D, Khayatnezhad M, and Taghavi M. Investigation of phosphoric acid fuel cell, linear Fresnel solar reflector and Organic Rankine Cycle polygeneration energy system in different climatic conditions. *Process Saf Environ Prot* 2021;147:993–1008.

[17] Wang Z, Zhang X, and Rezazadeh A. Hydrogen fuel and electricity generation from a new hybrid energy system based on wind and solar energies and alkaline fuel cell. *Energy Reports* 2021;7:2594–604.

[18] Ferriday TB and Middleton PH. Alkaline fuel cell technology—A review. *Int J Hydrogen Energy* 2021;46:18489–510.

[19] Alaswad A, Omran A, Sodre JR., *et al.* Technical and commercial challenges of proton-exchange membrane (PEM) fuel cells. *Energies* 2020;14:144.

[20] Meng X, Ren H, Yang X, Tao T, and Shao Z. Experimental study of key operating parameters effects on the characteristics of proton exchange membrane fuel cell with anode recirculation. *Energy Convers Manag* 2022; 256:115394.

[21] Bai F, Lei L, Zhang Z., *et al.* Application of similarity theory in modeling the output characteristics of proton exchange membrane fuel cell. *Int J Hydrogen Energy* 2021;46:36940–53.

[22] Taner T. The novel and innovative design with using H2 fuel of PEM fuel cell: Efficiency of thermodynamic analyze. *Fuel* 2021;302:121109.

[23] Olabi AG, Wilberforce T, Alanazi A., *et al.* Novel trends in proton exchange membrane fuel cells. *Energies* 2022;15:4949.

[24] Zhang X, Zhang T, Chen H, and Cao Y. A review of online electrochemical diagnostic methods of on-board proton exchange membrane fuel cells. *Appl Energy* 2021;286:116481.

[25] Priya K, Sathishkumar K, and Rajasekar N. A comprehensive review on parameter estimation techniques for Proton Exchange Membrane fuel cell modelling. *Renew Sustain Energy Rev* 2018;93:121–44. https://doi.org/10.1016/j.rser.2018.05.017.

[26] Mitra U, Arya A, and Gupta S. A comprehensive and comparative review on parameter estimation methods for modelling proton exchange membrane fuel cell. *Fuel* 2023;335:127080.

[27] Alizadeh M and Torabi F. Precise PEM fuel cell parameter extraction based on a self-consistent model and SCCSA optimization algorithm. *Energy Convers Manag* 2021;229:113777.

[28] Chen Y, Pi D, Wang B, Chen J, and Xu Y. Bi-subgroup optimization algorithm for parameter estimation of a PEMFC model. *Expert Syst Appl* 2022; 196:116646.

[29] Menesy AS, Sultan HM, Selim A, Ashmawy MG, and Kamel S. Developing and applying Chaotic Harris Hawks optimization technique for extracting parameters of several proton exchange membrane fuel cell stacks. *IEEE Access* 2019;8:1146–59.

[30] Pan M, Li C, Liao J., *et al.* Design and modeling of PEM fuel cell based on different flow fields. *Energy* 2020;207:118331. https://doi.org/10.1016/j.energy.2020.118331.

[31] Restrepo C, Garcia G, Calvente J, Giral R, and Martínez-Salamero L. Static and dynamic current-voltage modeling of a proton exchange membrane fuel cell using an input-output diffusive approach. *IEEE Trans Ind Electron* 2016; 63:1003–15. https://doi.org/10.1109/TIE.2015.2480383.

[32] Ćalasan M, Abdel Aleem SHE, Hasanien HM, Alaas ZM, and Ali ZM. An innovative approach for mathematical modeling and parameter estimation of PEM fuel cells based on iterative Lambert W function. *Energy* 2023; 264:126165. https://doi.org/https://doi.org/10.1016/j.energy.2022.126165.

[33] Mann RF, Amphlett JC, Hooper MAI, Jensen HM, Peppley BA, and Roberge PR. Development and application of a generalized steady-state electrochemical model for a PEM fuel cell. *J Power Sources* 2000; 86:173–80. https://doi.org/10.1016/S0378-7753(99)00484-X.

[34] Yang B, Wang J, Yu L., *et al.* A critical survey on proton exchange membrane fuel cell parameter estimation using meta-heuristic algorithms. *J Clean Prod* 2020;265:121660.

[35] Gupta J, Nijhawan P, and Ganguli S. Optimal parameter estimation of PEM fuel cell using slime mould algorithm. *Int J Energy Res* 2021;45:14732–44. https://doi.org/10.1002/er.6750.

[36] Miao D, Chen W, Zhao W, and Demsas T. Parameter estimation of PEM fuel cells employing the hybrid grey wolf optimization method. *Energy* 2020; 193:116616. https://doi.org/10.1016/j.energy.2019.116616.

[37] Singla MK, Nijhawan P, and Oberoi AS. Parameter estimation of proton exchange membrane fuel cell using a novel meta-heuristic algorithm. *Environ Sci Pollut Res* 2021;28:34511–26. https://doi.org/10.1007/s11356-021-13097-0.

[38] Xu S, Wang Y, and Wang Z. Parameter estimation of proton exchange membrane fuel cells using eagle strategy based on JAYA algorithm and Nelder-Mead simplex method. *Energy* 2019;173:457–67. https://doi.org/10.1016/j.energy.2019.02.106.

[39] Deepanraj B, Gugulothu SK, Ramaraj R, Arthi M, and Saravanan R. Optimal parameter estimation of proton exchange membrane fuel cell using improved red fox optimizer for sustainable energy management. *J Clean Prod* 2022; 369:133385.

[40] Yao B and Hayati H. Model parameters estimation of a proton exchange membrane fuel cell using improved version of Archimedes optimization algorithm. *Energy Reports* 2021;7:5700–9. https://doi.org/10.1016/j.egyr.2021.08.177.

[41] Duan F, Chen C, Song F, Zhang S, Li X, and Eslami M. An optimal parameters estimation for the proton exchange membrane fuel cells based on

amended deer hunting optimization algorithm. *Sustain Energy Technol Assessments* 2023;58:103364. https://doi.org/10.1016/j.seta.2023.103364.

[42] Zaki Diab AA, Tolba MA, Abo El-Magd AG, Zaky MM, and El-Rifaie AM. Fuel cell parameters estimation via marine predators and political optimizers. *IEEE Access* 2020;8:166998–7018. https://doi.org/10.1109/ACCESS.2020.3021754.

[43] Abaza A, El-Sehiemy RA, Mahmoud K, Lehtonen M, and Darwish MMF. Optimal estimation of proton exchange membrane fuel cells parameter based on coyote optimization algorithm. *Appl Sci* 2021;11:2052.

[44] Houssein EH, Hashim FA, Ferahtia S, and Rezk H. An efficient modified artificial electric field algorithm for solving optimization problems and parameter estimation of fuel cell. *Int J Energy Res* 2021;45:20199–218. https://doi.org/10.1002/er.7103.

[45] Fathy A, Abdel Aleem SHE, and Rezk H. A novel approach for PEM fuel cell parameter estimation using LSHADE-EpSin optimization algorithm. *Int J Energy Res* 2021;45:6922–42. https://doi.org/10.1002/er.6282.

[46] Ashraf H, Abdellatif SO, Elkholy MM, and El-Fergany AA. Computational techniques based on artificial intelligence for extracting optimal parameters of PEMFCs: Survey and insights. *Arch Comput Methods Eng* 2022;29:3943–72.

[47] Zhang Y, Huang C, Huang H, and Wu J. Multiple learning neural network algorithm for parameter estimation of proton exchange membrane fuel cell models. *Green Energy Intell Transp* 2023;2:100040. https://doi.org/10.1016/j.geits.2022.100040.

[48] Kartal F and Özveren U. Investigation of an integrated circulating fluidized bed gasifier/steam turbine/proton exchange membrane (PEM) fuel cell system for torrefied biomass and modeling with artificial intelligence approach. *Energy Convers Manag* 2022;263:115718. https://doi.org/10.1016/j.enconman.2022.115718.

[49] Hachana O. Accurate PEM fuel cells parameters estimation using hybrid artificial bee colony differential evolution shuffled complex optimizer. *Int J Energy Res* 2022;46:6383–405. https://doi.org/10.1002/er.7576.

[50] Aykut Korkmaz S, Ayten Çetinkaya S, Yuksel O, Konur O, Emrah Erginer K, and Ozgur Colpan C. Comparison of various metaheuristic algorithms to extract the optimal PEMFC modeling parameters. *Int J Hydrogen Energy* 2024; 51:1402–1420. https://doi.org/10.1016/j.ijhydene.2023.05.347.

[51] Shaheen A, El-Sehiemy R, El-Fergany A, and Ginidi A. Fuel-cell parameter estimation based on improved gorilla troops technique. *Sci Rep* 2023;13:1–17. https://doi.org/10.1038/s41598-023-35581-y.

[52] Huo W, Li W, Zhang Z, Sun C, Zhou F, and Gong G. Performance prediction of proton-exchange membrane fuel cell based on convolutional neural network and random forest feature selection. *Energy Convers Manag* 2021; 243:114367. https://doi.org/10.1016/j.enconman.2021.114367.

[53] Ma J, Liu X, Zou X., *et al.* Degradation prognosis for proton exchange membrane fuel cell based on hybrid transfer learning and intercell differences. *ISA Trans* 2021;113:149–65. https://doi.org/10.1016/j.isatra.2020.06.005.

Chapter 6

High-power charging solutions for electric ships

Adel El-Shahat[1], Zain K. Mirza[1], Jason Yu-Chen Qian[1] and Ziad M. Ali[2]

Electrification is one of the several ways to help lower greenhouse gas (GHG) emissions and help reduce the effects of climate change. The International Maritime Organization (IMO) has adopted several measures to reduce GHG emissions from international shipping by adopting electrification into the shipping system and an Energy Efficiency Design Index (EEDI) for new ships that include energy management plans referred to as a ship energy efficiency management plan (SEEMP). In the past, ships have used alternating current (AC) but there has been a shift to using direct current (DC) due to several reasons including higher power densities, more efficient integration of alternative fuel types, higher power quality, higher efficiencies and lower weight and area needed by DC equipment in comparison to AC equipment – 56% and 41%, respectively.

6.1 Implementation of BiDC grids on ships generation

DC grids can have unipolar (UDC) or bipolar (BiDC) power system. A BiDC grid can have either two positive and negative poles, OR three conductors (positive, negative, and natural). Figure 6.1 shows a method in which a BiDC grid is adapted. The converters are connected between a single pole and the neutral with the drivers from the thruster and propulsion being connected between the positive and negative poles. Figure 6.2 shows a different way to create a BiDC grid using a three-phase synchronous generator. This does create issues of added weight and costs but can be avoided by adding a dual stator synchronous generator.

[1]School of Engineering Technology, Purdue University, USA
[2]College of Engineering at Wadi Addawasir, Prince Sattam Bin Abdulaziz University, KSA

Table 6.1 compares several different types of BiDC generation methods in terms of several different parameters.

Harmonics is an issue to take into consideration when there are power imbalances between a pole and the neutral that can lead to higher copper losses and

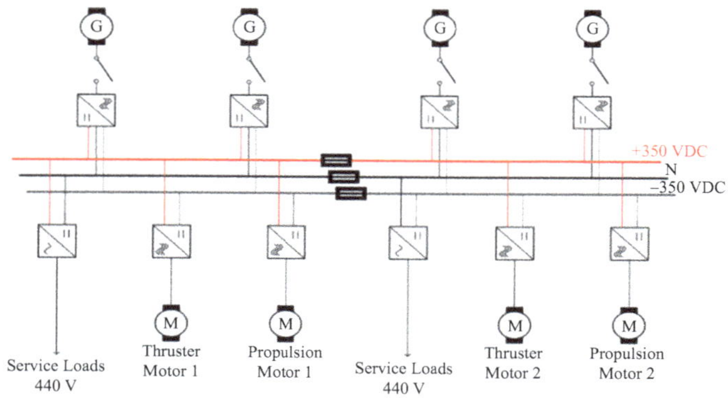

Figure 6.1 Platform supply vessel BiDC power system

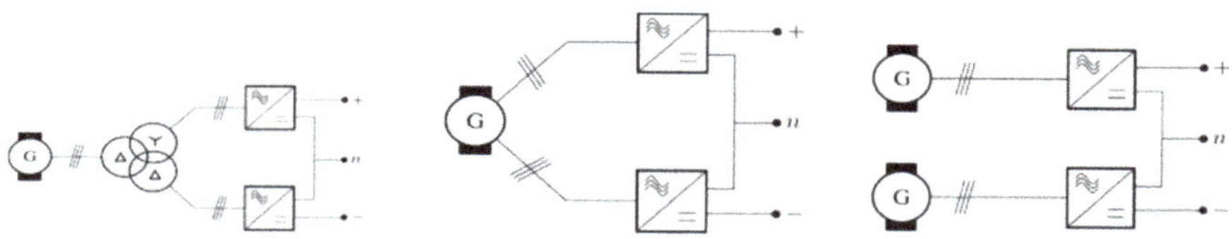

Figure 6.2 Ways to create BiDC grids on ships

Table 6.1 BiDC generation methods comparison

Parameter	**Single synchronous generator**	**Dual stator generator**	**Dual synchronous generator**
Cost	Low	Medium	High
Space	Medium	Low	High
Weight	High	Low	Medium
Fault tolerance	Low	Low	High

Table 6.2 Power balancing converters comparison

Balancing converter	Switch voltage	Passive component amount	Symmetric switch loading
Buck/boost	High	Low	Yes
Cuk	Low	High	Yes
Sepic	High	High	No
Zeta	High	High	No
Three-level converter	Low	Low	Yes
B6 converter	Low	High	Yes

torque pulsations that can lead to catastrophic incidents. This can be fixed by using controlled or active rectifiers.

6.1.1 Power balancing

Power imbalance is inherent with BiDC grids; the unbalanced power flow between poles and neutral may lead to voltage shifts of the neural line which can trip critical equipment and result in lowered efficiency. There are several ways to achieve power balancing:

1. Three-level converters can be controlled to supply the required power to the different poles.
2. The power converters can be controlled in a synchronous reference frame to achieve balance.
3. Application of DC–DC converters on the DC grid size.

Table 6.2 provides a summary of different converters that can be used in power balancing.

6.2 Challenges of BiDC implementation

6.2.1 Stability

BiDC ships would consist of several power electronics converters that can bring stability issues in the grid. Power electronic converters are used to interface prime movers as the inertia of the grid decreases. Droop control and virtual synchronous generator algorithms can be used to help mitigate these issues, but still, there is considerable research to be done to understand these issues on ship's power systems.

Power grids on ships are weak due to the installed generation capacity not being significantly higher than the highest total load demand. Weak grids are known to be susceptible to becoming easily unstable when there are changes in load requirements.

6.2.2 Protection

Protection is very important for the power grid which can be more challenging for DC grids – more than an AC grid. These challenges include:

1. DC grid does not have zero current crossings so that when a fault occurs in a DC grid the current does not decrease and an electric arc forms when contacts are trying to open. This may lead to the erosion of the circuit breaker's contacts which may lead to catastrophic situations in some cases.
2. DC grids are fed with power electronic converters, when faults occur the converters will shut down in order to break the supply for the grid. This makes a limited selectivity in the detection and isolation of faults.
3. If there is a fault on the DC side, the filter capacitor and cable capacitance will dump all their energy into the fault in a very short period of time leading to a higher rate of rise current during the fault.
4. The power electronic converters used in DC grids do not provide sustained high currents when a fault occurs; this leads to problems in detection and selectivity isolation.

There are several methods for designing protection systems for DC microgrids. The first method is based on de-energizing the whole grid – where the main converter will be turned off and the current is reduced. When the current is dropped below a certain value, the contactors are operated to isolate the faulty power supplies. Then, the bus will be reenergized, Figure 6.3 illustrates this process with a graph of how the current changes with time.

Another method of DC grid protection is given by the NPR 9090:2018 standard [1]. In DC grids, some of the power sources may have very high short circuit currents and may portray risks for other equipment in the system. In this case, the standard and generation and load units are distributed in different zones based on their fault characteristics – short circuit, nominal current limits, and current flow direction.

Grounding is a critical aspect of power systems. Selecting inappropriate grounding methods can lead to corrosion issues in ships – and corrosion in DC

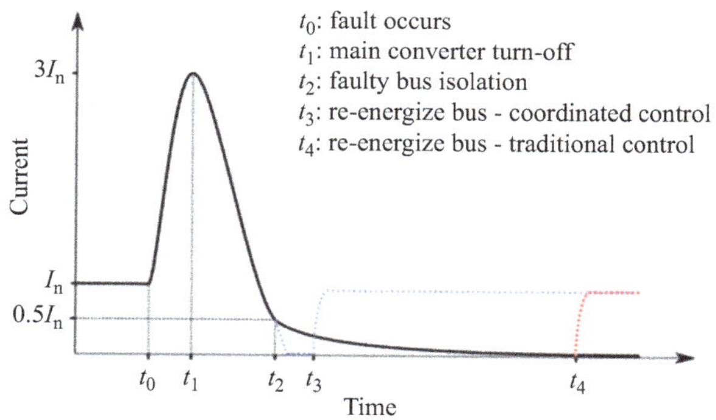

Figure 6.3 De-energizing protection scheme for DC grids

systems is more severe than AC systems. This is due to the non-oscillatory current and voltage magnitude in DC systems. The grounding of negative terminals leads to the corrosion of metal parts which happens because of the electrochemical reactions that take place on the metal surface. These corrosion concerns lead to higher operation and maintenance costs as the corrosion is not always visible to the naked eye.

When there are multiple points of grounding in BiDC grids, there is a possibility of the flow of circulating currents between them. This happens because of the voltage unbalance in the grid, which leads to the different potential of the neutral line at various nodes. The extent of the effects is dependent upon the type of grounding connection and the grid architecture – which may lead to an increase in corrosion. This is why designing a proper grounding scheme is essential to ensure the safe operation of a ship.

6.3 Standardization

BiDC grids are still a relatively new technology with limited real-world applications. The limited adoption of BiDC technology is due to the limited number of standards. The Netherlands has published guidelines for LVDC grids. In this standard, the DC system is segregated into several safety zones with each zone having its own risks and measures for protection. The lack of standardization makes it difficult for manufacturers to build devices suitable for the purpose.

6.4 Conclusion

The industry is gradually shifting to the concept of more electric ships to increase the efficiency of operation and the reduction of GHG emissions – with BiDC grids having advantages compared to three-phase AC and UDC grids. The methods and challenges of implementing BiDC grids on ships were summarized in this source. There are different methods to form BiDC grids on ships, the choices made in the design and technology selection are determined by the power quality requirements, costs, and space. Balancing converters are also an essential part of the BiDC grid and the type chosen can vary depending on the size of ships.

Even though there seem to be several advantages to the implementation of BiDC grids on ships, there are still several challenges associated with it. Ascertaining the stability of operation when there is a large amount of power electronic components are used. The parity between the total installed generation and load capacity can lead to weakening leading and instability. Also, the biggest challenge is the inability of cost-effective and reliable protection systems for BiDC grids. The lack of standards for LV BiDC grids also creates a challenge – the availability of such standards would lead to greater adoption and acceptance of the technology in the market. These challenges need to be addressed first to make BiDC grids feasible on ships.

6.5 Introduction: smart grid and DC microgrid technologies into maritime electrical networks

Maritime power systems have greatly developed with the increase in demand of electrical loads. With regulations emissions from governments and fuel efficiency requirements on the rise, it asks for more efficient ships and the use of all-electric ships (AES) as opposed to the conventional mechanical propulsion system. AES uses the concept of integrated power systems (IPS) that minimize the number of generators used in the ship by using intelligent methods for meeting load demands and needs. The shipboard IPS can be regarded as a large-scale onboard microgrid with specific requirements. Current research shows that IPS is turning to DC power systems, especially in advanced control, management, and optimization methods.

6.5.1 DC power architecture

The world's first cruise vessel with an electric propulsion system, *Queen Elizabeth II*, is an example of an existing AC shipboard power system. Figure 6.4 shows the power architecture of the power system. The electric power plant is connected with the vessel's main bus which drives the two major electric propulsion systems. The loads are powered through transformers and power electric converters. The conversion to a diesel-electric power system was expected to improve the fuel efficiency up to 35%. Research has shown the inadequacies of the AC power architecture which include:

- Generators set shave to work in fixed peed, limiting the further improvement in fuel efficiency.

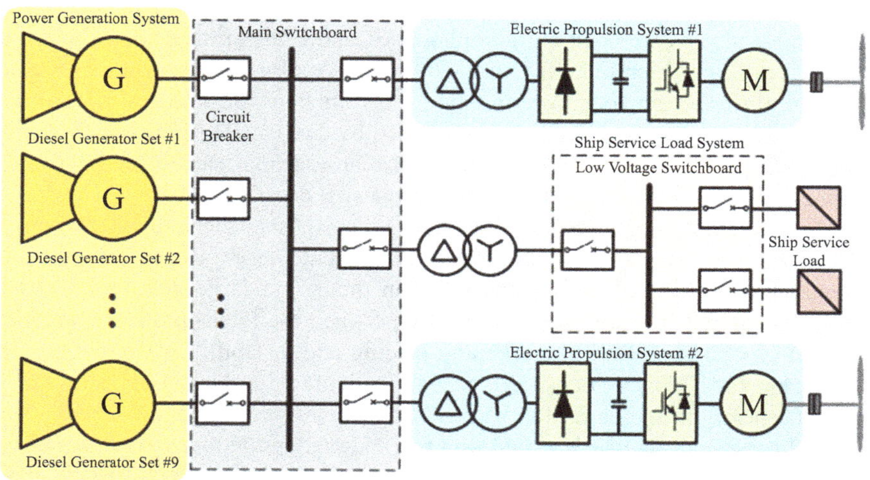

Figure 6.4 AC power system architecture of Queen Elizabeth II

- Unwanted reactive power flow and power quality problems from AC power architecture
- Space and weight issues from the transformers used.
- Potential risks of system disintegration

Taking the issues mentioned above into consideration, there have been trends toward the dc power system. The dc power system has several major advantages which include:

- Using compact power electronic converters instead of bulky transformers
- Easier parallel connection or disconnection for dc power sources
- Elimination of harmonic and imbalance problems
- Elimination of synchronization problems
- Elimination of reactive power flow

Furthermore, DC-based IPS brings a wide range of advantages when considering the specific needs of shipboard power systems. DC power architectures will eliminate the bulky transformers and reduce the rating of switchgear, decreasing weight and space issues that AC power systems have. DC power architecture could also provide better survivability and limitation of fault current. Since there are higher power levels required in AES application, Figure 6.5 shows the only available design option for DC-based IPS.

In Figure 6.5(a), the renewable energy sources (RESs), energy storage systems (ESSs), and local electrical loads are packed together with the DC bus in order to enable islanding operation, making the system fully resistant to major blackouts in the main grid. Elimination of reactive power and synchronization problems makes the whole system simpler to design, control, and coordinate. Figure 6.5(b) shows a three-wire bipolar-type DC power distribution system which has inherent redundancy and higher reliability.

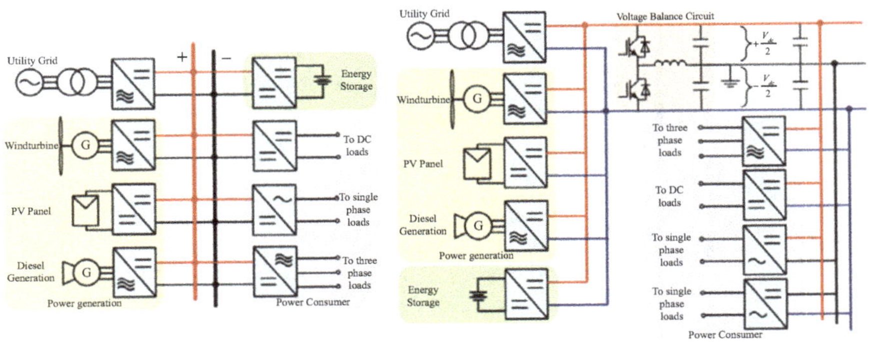

(a) (b)

Figure 6.5 Typical power architecture of single DC microgrids: (a) common architecture and (b) bipolar architecture

6.5.2 Control and coordination of microgrid-based power system

Even though there are several benefits offered by DC-based IPS, achieving voltage regulation can be challenging. According to IEEE Std 1709-2010, the shipboard DC power system needs to fulfill the following control objectives:

1. Power system stability: Able to maintain autonomous equilibrium in normal conditions and regain a state of operating equilibrium after being subjected to a physical disturbance.
2. Power quality: Able to maintain or restore the common dc buses at their nominal voltage with acceptable voltage tolerance.
3. Power management: Able to optimize systemic efficiency by intentional scheduling or intervention without affecting maximization of power supply to the demand side.

Research in advanced control and coordination techniques has been investigated for DC microgrids; to achieve different control functions, hierarchical control scheme is proposed with these defined levels, which can also be seen in Figure 6.6:

1. Level 0 (inner control loops): The fundamental control loops to regulate the output voltage and/or current within each power electronic converter connected to the microgrid.
2. Level 1 (primary control): The control methods to emulate the physical behaviors that make the system stable and more damped power sharing.

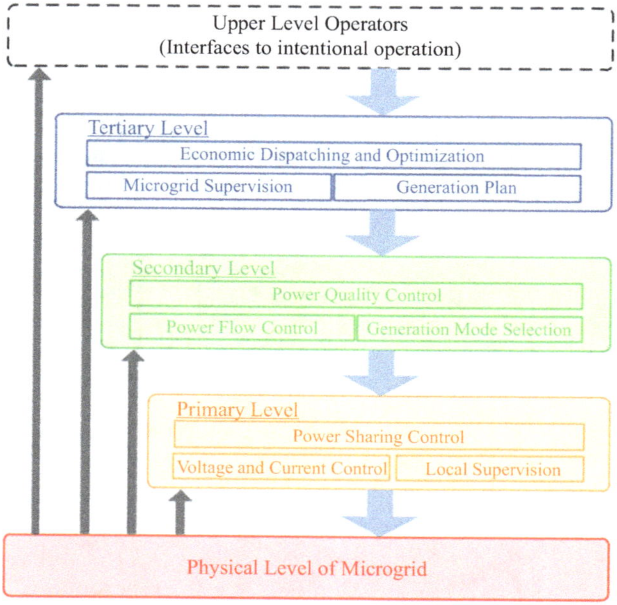

Figure 6.6 Different control levels in hierarchical control

3. Level 2 (secondary control): The control methods to ensure the major variables of the system are within the required values.
4. Level 3 (initial tertiary control): The control methods to manage and control the power flow among the upper layer grid and/or other microgrids.
5. Upper levels (extended tertiary control): The control and decision-making methods to achieve extra targets (practical economic benefits, etc.).

Figure 6.7 shows a control architecture applying hierarchical control in a generalized DC microgrid. Droop control is the primary control method for active power sharing and in recent studies, it has been shown that either output power or output current can be selected as the feedback signal in the droop control. The droop coefficient is regarded as a virtual internal resistance, with the droop control consisting of the physical connection of DC sources and simplifying the design of the parallel converter systems in the microgrid. A small voltage deviation is introduced by droop-based primary control and the secondary control is introduced to compensate the voltage deviation. In large-scale DC microgrids, hierarchical control is the preferred choice because it allows decoupling behavior between different control layers. Hierarchical control is achieved by using local control of the power electronic interfaces while at the same time coordinating the control of all these components. The secondary and tertiary control levels are relying on the cooperation of several or all local controllers, and this is why the coordination in the microgrid will impact the system's overall stability, reliability, and performance. Coordination methods can be divided into three categories which are centralized, decentralized, and distributed.

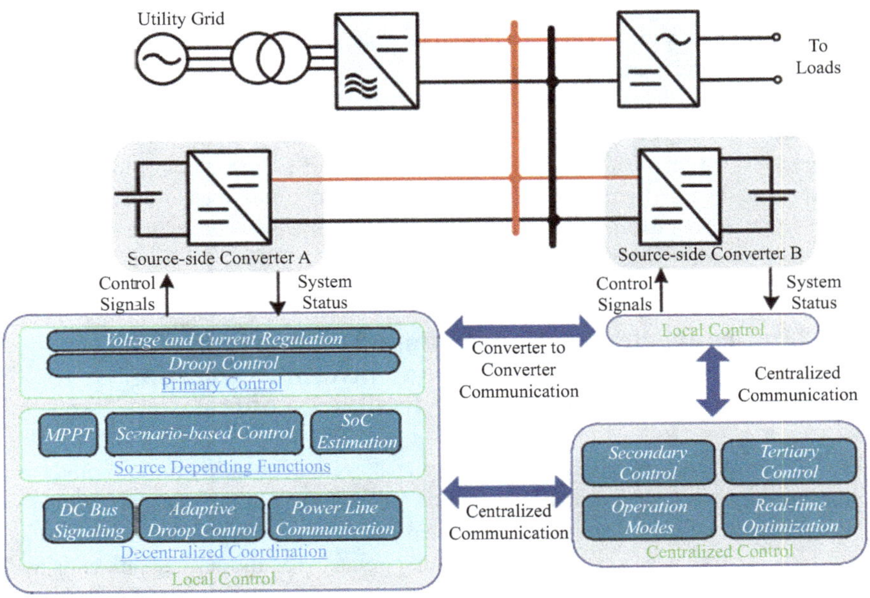

Figure 6.7 Hierarchical control in practical DC microgrids

6.5.2.1 Smart grid technologies, the key to smart onboard power systems

The common trend for the future of power systems is evolving for higher intelligence and efficiency. Intelligent management and smart protection are being introduced to the microgrid as an extension to existing systems to achieve goals of higher efficiency and fewer emissions.

6.5.2.2 Smart coordinated management for lower cost and less emission

The IMO has been making efforts in recent years to promote SEEMP to encourage emission limitations. When designing a SEEMP, it is possible to incorporate advanced offline optimization algorithms to improve the fuel-saving effect. To maximize fuel efficiency and in turn minimize emissions a method to combine scenario-based multi-mode control and real-time optimization – this can be done with the different constraints given by the different levels of control and the detailed system status. Implementing SEEMP requires joint management in the generation and demand side. In terms of generation, the DC distribution system will allow each prime mover to operate in a variable speed mode independently without synchronization limits. The generation side management will keep the onboard generators whether working in high-efficiency area or at idle speed.

6.5.2.3 Smart protection and reconfiguration for a fault-tolerant and highly reliable system

Protection of DC power systems is an important and challenging task that requires plans that incorporate complex coordination between power converters and protective functions. Additionally, the instantaneous overcurrent capability of power electronic converters must be limited in order to avoid equipment damage. As a result of this shipboard, IPS may be overdesigned in terms of power electronic equipment – which also gives arise to space constraints.

Research on enabling protective functions using power electronic converters has been made, but there are communication delays and measurement failures. Studies have shown that measurement failures could be overcome by using outliner data detection and reconstruction algorithms.

6.6 Multiconverter medium-voltage DC power systems on ships: constant-power loads instability solution using linearization via state feedback control

Future medium-voltage DC (MVDC) power systems for ships must include bus voltage stability which is a major concern. Voltage instability may be brought on by the presence of high-bandwidth regulated load converters (constant power load, CPL). An approach to control design that begins with modeling and ends with control implementation is provided in this source. To combat the CPL's destabilizing influence and guarantee the stability of the MVDC bus voltage, a control solution based on linearization via state feedback (LSF) is developed. A new comprehensive model that can accurately depict the overall behavior of a second-order nonlinear differential equation is used to study a multiconverter shipboard DC grid.

LSF approach uses the DC–DC converters that connect power sources to the bus to take advantage of system nonlinearities and create a linear system. After that, the appropriate pole placement may be achieved by using conventional linear control techniques. A sensitivity analysis is used to verify the LSF control design with reference to the system parameter mismatch and to assess the likelihood of an over-linearization strategy. By using a two-way comparison and time-domain numerical simulations, the suggested control is validated in the presence of relevant disturbances.

6.6.1 Introduction

The most recent AES projects use medium-voltage DC (MVDC) power systems as its foundation. On this subject, a recent IEEE standards have been released that list a number of MVDC grid characteristics that need in-depth research, including stability study. In these systems, short-distance direct connections between power converters and the common MVDC bus are used to supply power to loads.

A constant power load (CPL) characteristic may appear when broad control bandwidths are used for load converters. A CPL has negative incremental resistance, which has a tendency to cause instability. System stability must thus be carefully assessed. There are several contributions to the literature on CPL instability that provide a variety of solutions to the issue. In this study, a thorough methodology is presented, starting with modeling and ending with control implementation. The suggested solution's robustness is also measured. A novel comprehensive model that may accurately depict the overall behavior of a second-order nonlinear differential equation is suggested after an analysis of a multiconverter structure. This model is then used for the implementation of a LSF, which can be seen in Figure 6.8.

The suggested method in Figure 6.8 aims to contribute to the design of a shipboard DC grid's full voltage management system, starting with data on plant components (i.e., generators, loads, converters, filters, cables). Some of these data (generators, loads, shipboard cables) are taken for granted as inputs, while others (DC–DC converters, filters) must be accurately established using the design process outlined in this work.

The source provides and details for multiconverter MVDC system modeling which is divided into four sections including:

(A) Power system topology
(B) MVDC circuit model
(C) MVDC simplified circuit model
(D) Equivalent second-order model

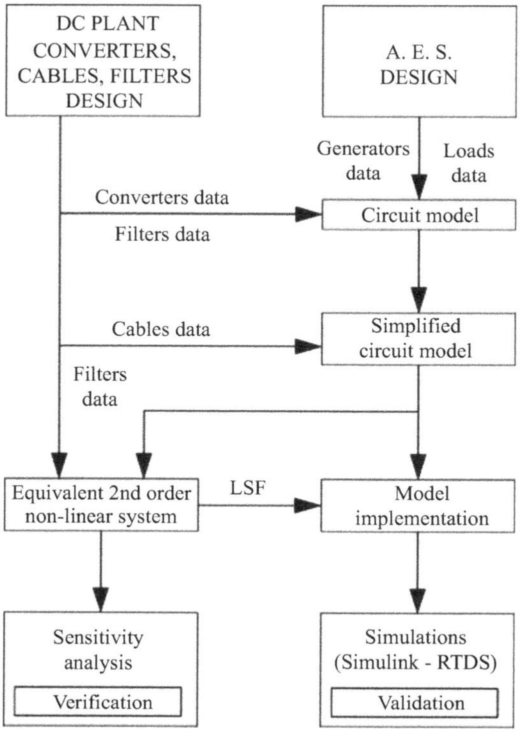

Figure 6.8 Voltage control design procedure for shipboard MVDC grids [2]

The sections also provide formulas that are used and the respective circuitry. The source also provides detailed information with formulas on *LSF-Based MVDC Bus Voltage Control, Filters Design*, and *LSF Sensitivity Analysis*. The last two sections of the source which are *Model Implementation* and *Simulation Results* will be discussed in detail below.

6.6.1.1 Model implementation

Two simulation setups – the AVM and the detailed switching model – are utilized to construct the LSF control approach. However, a thorough model that offers switching quantities verifies the veracity of simulation findings.

Average value model implementation

The block schematic shown in Figure 6.9 shows the AVM system implementation which is described as follows. DC–DC converters may be modeled using the gain value, supposing that the output voltages of the diode converters remain constant during load supply. Therefore, these DC–DC power converters are viewed as "voltage actuators" with little dynamics. An outer integral controller, which analyzes the bus voltage feedback and produces the operating point command for the

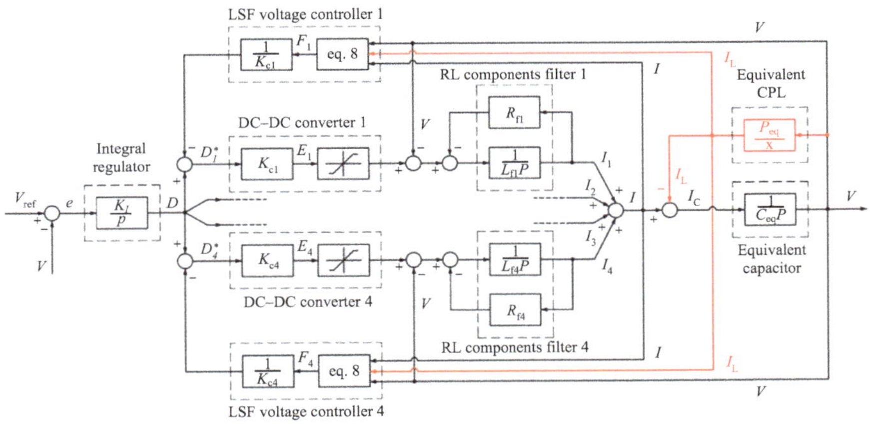

Figure 6.9 Block scheme of the multiconverter MVDC system [2]

converters duty cycle, is responsible for bringing the multiconverter MVDC system's voltage operating point to the desired level. Regarding the quick action of the inner control loops, this integral control loop is built with sluggish dynamics. The LSF voltage controller blocks carry out control operations, and it is their output commands that are combined to produce the duty cycle commands. The red lines seen in Figure 6.9 show the CPL nonlinear parts that are eventually canceled out. The load sharing of the generators achieved by the contributions may be impacted by steady-state errors in the case of an approximation of the time constant that is not precisely defined. In this situation, additional generator current control loops could be used to regulate the load sharing coefficients in order to ensure the correct steady state power sharing.

Detailed model implementation

Simulink, which is based on the state equation simulation approach and is often used for offline open- and closed-loop control algorithm simulation studies, is used to construct the system AVM and control methods. It goes without saying that the effective implementation of this aids in the evaluation of control techniques. However, it was unable to support the switching-level performance analysis of power converters, which included determining whether the filter voltage and current ripple complied with design requirements and whether the power converter switching frequency had an impact on the regulated natural frequency.

In order to easily provide access to real-time hardware in the loop online testing, it is desirable to have a real-time detailed model implementation in Real-Time Digital Simulator (RTDS) that is based on another simulation technique (the resistive companion simulation method). In addition to correcting the aforementioned flaws, this real-time comprehensive model implementation would significantly increase the realism of the MVDC system model. An ideal DC voltage source coupled to a DC–DC converter represents a comprehensive generating system that replaces the standard unit. The converter's output filter is the second-order RLC filter. These four intricate generating systems, whose switching frequency has been set at 1.5 kHz, are used to test and validate the corresponding ideal CPL algorithm.

6.6.1.2 Simulation results

The validity of the models is implemented and tested using appropriate case studies that are detailed in this section of the source. A sudden generating system disconnect represents the worst-case situation for shipboard power systems; hence, it has been chosen as the test case for the LSF control approach. A multiconverter MVDC system with the structure shown in Figure 6.10 and characteristics shown in Table 6.3 is supposed to provide a CPL load of 18.5 MW using three DC producing systems (B1, B2, and B3) with a combined rated power of 42 MW at the first steady state. At steady state, the operating point is maintained while feeding the CPL.

To imitate the loss of B3, CB3 opens at $t = 6$ s, making the total rated power 26.25 MW. The load is automatically re-shared among the two remaining DC-producing systems using the load-sharing coefficients that are automatically computed and adjusted. The output powers of the other converters are around 70% of their rated ones after B3 disconnection. While the control gains are chosen such that the outer voltage control loop results dynamically disconnected from the inner ones, the sluggish dynamics of the outer integral control loop are defined by an analogous time constant of 0.5 s.

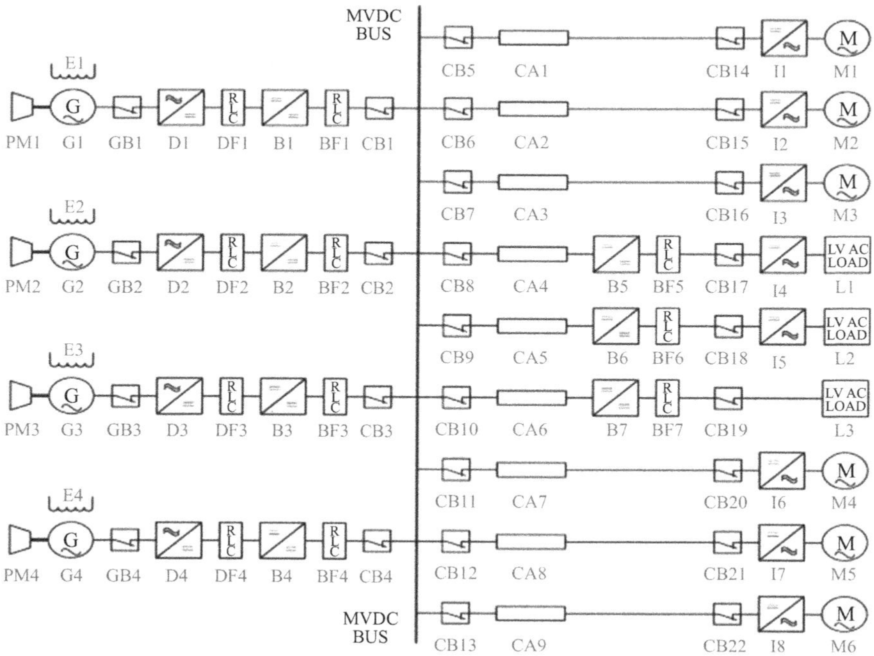

Figure 6.10 Multiconverter MVDC power system [2]

Table 6.3 Multiconverter MVDC system filter parameters [2]

	BF1/BF3	BF2/BF4
P_{nk} [MW]	15.75	10.50
V_{dnk} [V]	8,910	8,910
V_{nk} [V]	6,000	6,000
D_{nk}	0.67	0.67
I_{nk} [A]	2,494	1,662
f_{sk} [Hz]	1,500	1,500
$\Delta P_{\%k}$	5	5
$\Delta P_{\%k}$	3	3
$\Delta I_{\%k}$	30	30
$R^0{}_k$ [Ω]	2.3	3.4
R_{fk} [mΩ]	126.6	189.9
L_{fk} [mH]	1.7	2.6
C_{fk} [μF]	346.3	230.9
ω_{0fk} [rad/S]	1,250	1,250
ξ_{fk}	−0.48	−0.48

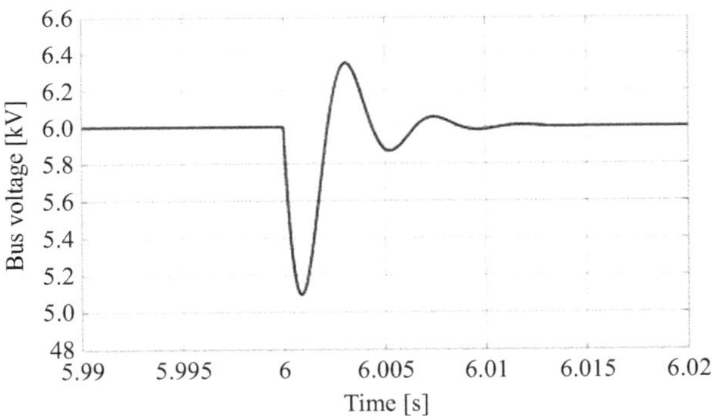

Figure 6.11 Bus voltage transient (fl on, fc on) [2]

Average value model results

The overall damping factor for the initial system, when f_l and f_c are disabled, is −0.32, which is indicative of an unstable system. The following comments discuss the beneficial impact of applying the control strategy. To demonstrate the power system's dynamic behavior following the disconnect of the producing system, three transients are illustrated in Figures 6.11–6.13 under the supposition that the linearizing function f_l and the control function f_c are both active. The voltages shown in Figures 6.11–6.13 exhibit a steady progression when under the control of the intended dynamic. However, the transient shown in Figure 6.14 demonstrates the need of maintaining the stability of the system: the uncompensated CPL nonlinearity generates large unstable oscillations which would necessitate the use of

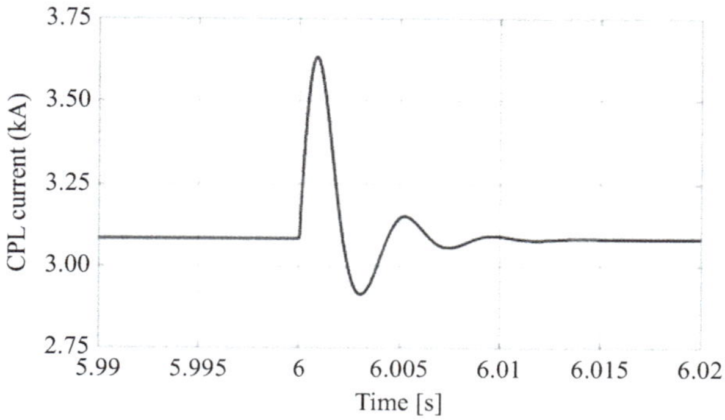

Figure 6.12 CPL current transient (fl on, on) [2]

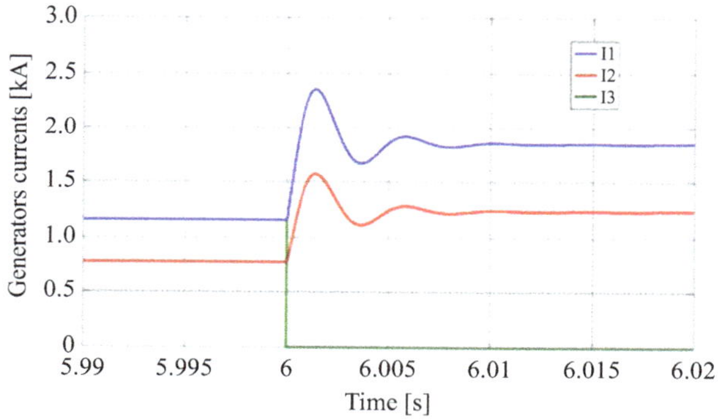

Figure 6.13 Generators currents transient (fl on, fc on) [2]

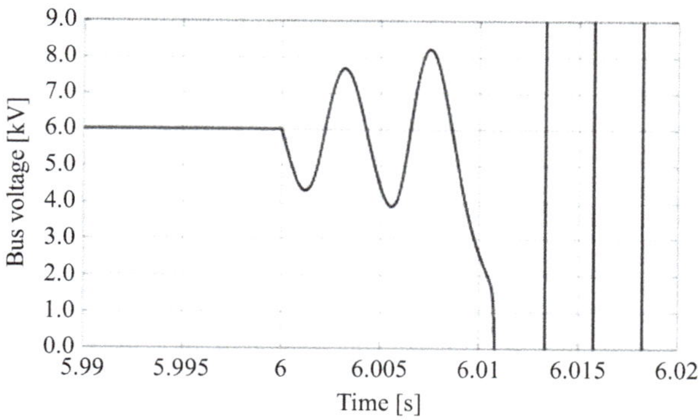

Figure 6.14 Bus voltage (fl off, fc on) [2]

voltage safeguards in a real plant. The sensitivity analysis proposed in V is used to confirm stability consequences resulting from mismatched system parameters in the worst-case scenario.

6.6.1.3 Detailed model design

Using a sophisticated model and RTDS, the identical transitory process of losing the primary generating system B3 is reproduced. The associated outcomes, which include Figures 6.15–6.17, show highly uniform transient character traits. On the one hand, the anticipated damping factor of 0.3 and natural frequency of 1500 rad/s are met, yielding the required transient control performance. On the other hand, the steady state bus voltage and generator current performance are within 3% tolerance, correspondingly, demonstrating accuracy of the filter design and 30% ripple constraint. The effective pairing of the RTDS (detailed model) and Simulink (AVM) simulation results demonstrate the viability of the suggested LSF control technique seen as sensible and efficient.

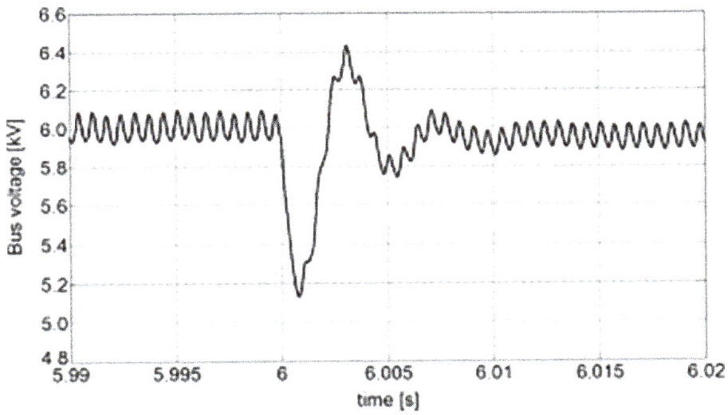

Figure 6.15 Bus voltage transient [2]

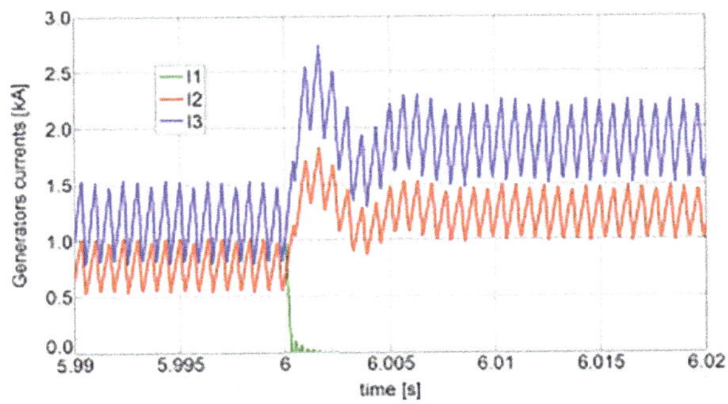

Figure 6.16 Generator's current transient [2]

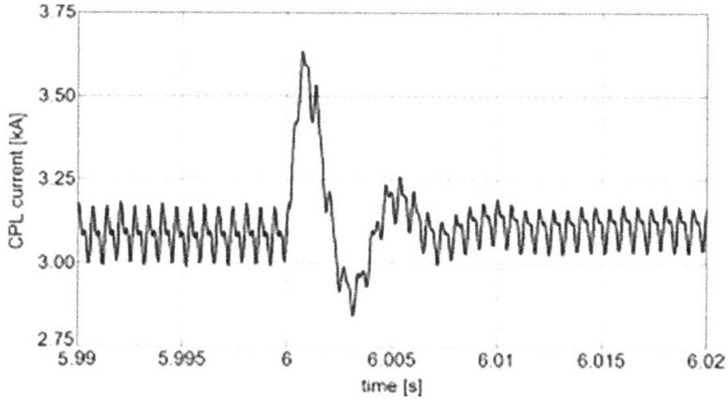

Figure 6.17 CPL current transient [2]

6.6.2 Conclusion

A voltage control design process for MVDC grids is provided in this study. The control strategy is based on state feedback linearization and aims to address the problem of CPL stability in a multiconverter MVDC system. The model of an MVDC power system, which is derived from an original, real-world AES AC power system, is subjected to the suggested control.

Feedback linearizing controllers have been extensively addressed in literature, as is well known. However, the current work offers a thorough voltage control design process for DC radial grids that benefits from a unique system model simplification. It has been shown that a complicated multiconverter DC system may be represented by an equivalent nonlinear second-order model given reasonable design assumptions. The LSF is used to make this system linear, after which conventional linear systems theory may be used to manage it with ease.

In order to satisfy the needs of the shipboard plant, implementation considerations have been incorporated into the proposed control architecture. Gained benefits are as follows: besides MVDC bus voltage control, dynamic power sharing is obtained among generating systems, which are responsible for providing load power and stabilizing power; the proposed control design, supplemented by suitable filters, is considered normally available from onboard measurements (generating converter currents, bus voltage, and load converter currents). A sensitivity (small signal) analysis has offered a design verification, assessing control resilience in case of parameter uncertainty, and stressing the most sensitive parameters (robustness can be augmented using over-linearization).

A two-way comparison using a system AVM created in the Simulink environment (single processor) and a real-time detailed switching model built in the RTDS environment has been used to evaluate the LSF-based voltage control architecture (multiprocessor). Time-domain transient responses (bus voltage, generator currents, and equivalent CPL current) in the presence of a pertinent perturbation (loss of one generator out of three functioning) have been used to validate the control action of the LSF method.

6.7 DC-grid system for ships: a study of benefits and technical considerations

AC has long served as the foundation for ships' main electric power systems. However, maritime engineers began to doubt the AC grid system's effectiveness, which they had previously taken for granted, and they looked for a more effective and environmentally friendly electric power distribution system. The DC system was used for the electrical distribution system in ships and coupled with the AC grid in line with this trend in the maritime industry. In this regard, this chapter outlines the advantages of the DC-grid system for maritime applications in terms of technology, economics, and the environment. Several types of ships that have either implemented or intend to implement the DC-grid system exist. The technical aspects of the fault protection topology, the power-sharing (balancing) topology, power quality/stability concerns, power source control strategies, the DC arc flash danger, and international regulations/standards for DC-grid ships are also discussed. An analysis of the future of the DC-grid system in ships follows the conclusion.

6.7.1 Introduction

Although the DC grid system has been around for a while, the AC grid system has been around for a very long period as well. Long-distance transmission of electric power was challenging though. As a result, the AC system – which employs transformers to efficiently transmit high voltages – was chosen as the industry standard.

The DC has made a reappearance in both land and maritime industries after taking a long hiatus. The expansion of the DC grid has been aided by advancements in power electronics, renewable energy, and ESSs. Additionally, the market's needs for cleaner, more effective electric power in opposition to stronger environmental rules have an impact. For instance, the GW-class DC system is garnering interest as a potential method for transmitting a significant quantity of electrical power. Because it has no reactive power losses, this HVDC (high-voltage direct current) transmission system can carry more electrical power over long distances than an equivalent AC system. Recent spacecraft, space stations, airplanes, electric cars, and ships have also utilized small-scale DC grids.

This report conducted an in-depth analysis of the DC-grid system to broaden its use in the maritime sector in response to the current surge of interest in research on the DC grid. First, the "benefits of DC-grid" section lists a number of important advantages the DC-grid system has over the AC system. Additionally, the "types of the DC-grid" section include a variety of ships that have either implemented or intend to use the DC-grid technology. Additionally, in the section under "technical considerations," significant technical factors that support the DC-grid system in ships are covered. The DC-grid system's potential application in the maritime sector is then discussed.

6.7.2 Benefits of DC grid

(I) Power stability and quality benefits

Voltage and frequency must both be tracked and managed in the AC-grid system in order to keep the electricity stable. However, because there are no interactions involving reactive power in the DC-grid system, the system control is only focused on voltage. Therefore, compared to an AC grid, a DC grid has an advantage in maintaining power stability. Additionally, synchronizing generators is less complicated than synchronizing an AC grid. When synchronizing more than two generators in an AC-grid system, the voltage, frequency, and phase angle must all be taken into account. But under the DC-grid system, the only thing that matters is the voltage.

The major problematic factor in terms of the AC-power grid's quality is harmonic distortion. Harmonic distortion has been a problem for electric propulsion ships in particular, especially those with several variable frequency drives (VFDs). An inverter recreates the AC at the necessary speed for the road after a rectifier in a VFD transforms an AC to a constant voltage; this may cause harmonic distortion problems.

In the past, harmonic filters had been added to the installation to address this issue. Subsequently, a phase-shifting transformer was often built in front of the VFD to produce 12 or 24 pulses. Although an active front-end converter or other modern VFD may considerably minimize harmonics, their cost is somewhat greater than alternatives.

However, with the DC-grid system, converting an AC to constant voltage (DC) is not required; hence, the rectifier component of the VFD might be removed. By doing this, harmonic distortions and power losses into the connected load might be minimized.

As a result, ships with numerous VFDs for controlling motors like large lifting cranes, propulsion or thruster motors, pumps, compressors are ideally suited for the DC grid. To sum up, the DC-grid system offers the following advantages over the AC grid system in terms of power quality and stability:

- Reactive power freedom (increased power stability)
- Frequency freedom (easier synchronization of power sources)
- DC-based power distribution (which improves power quality and lowers harmonic distortions)

(II) Economic and environmental benefits

A genset (generator and prime mover) must operate in the AC-grid system at a specified fixed frequency of 50 or 60 Hz. However, a generator set's specific fuel oil consumption (SFOC) varies depending on the load, and it is normally optimized for a load factor of between 75% and 85% (ABB 2017a). As seen in Figure 6.18, the SFOC increases when the load is low.

The DC-grid system, on the other hand, allows gensets to run at various frequencies, giving it a larger working window and good fuel economy. Depending on

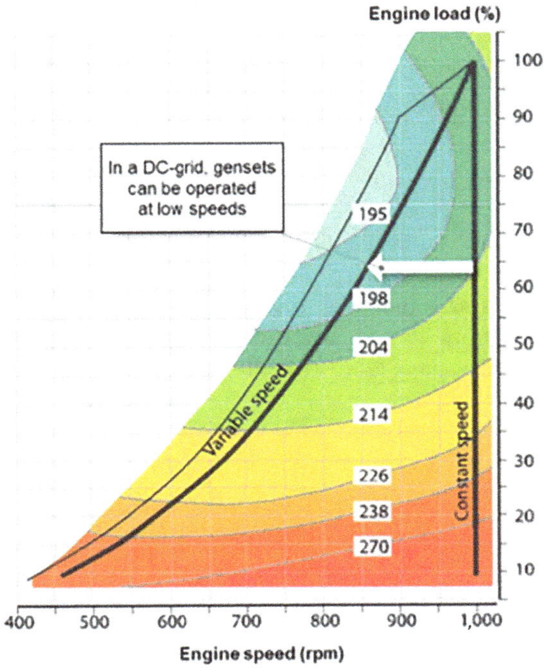

Figure 6.18 Example of a generator fuel consumption graph

the kind of ship, it is known that the DC-grid system might save fuel use and emissions by up to 20%.

As was already indicated, most electric propulsion ships contain extra equipment to reduce harmonic distortions, such as phase-shifting transformers or harmonic filters, which are no longer required for the DC grid. The rectifier component (AC/DC) of the VFD can also be taken out. More cargo areas can be created on a ship as a result.

Additionally, it is simple to include DC power sources into a DC bus such as fuel cells, lithium-ion batteries, and supercapacitors. The ESS in particular might be employed for a number of tasks, including peak shaving, load balancing, and absorbing regenerative power. As a result, the ESS can shorten genset runtime and increase energy effectiveness. Additionally, it can help lower the cost of genset maintenance. To sum up, the DC-grid system offers the following advantages over the AC grid in terms of economic and environmental benefits:

- Gensets' variable speed/frequency operation (reducing fuel consumption, emissions)
- Removing a rectifier for motor controls from a VFD
- Disposing of harmonic mitigation apparatus

- Simple DC power source integration with a DC bus (improving energy efficiency)
- It is not essential to use a separate tiny generator for harbor's power supply in the event of MW-class massive ships (saving the cost of a small genset)

The DC-grid system would be superior to two situations, when all advantages are considered. One example is the use of DC power sources as the main power source in ships. Additionally, ferries, cruises, and small feeder vessels may be the preferable ship types in this situation. Additionally, due to their cheap maintenance costs, autonomous or unmanned ships, like the planned semi-autonomous ferries from Fjord1 and the autonomous container feeder vessel (120 TEU) from YARA, have a high likelihood of installing a full-battery power system with the DC grid.

The ship that needs variable speed motors with VFDs for things like dynamic positioning (DP) thrusters, pump motors, or crane motors is the other situation. Offshore service boats, platform supply vessels, multipurpose support vessels, anchor handling tug supply vessels, research vessels, shuttle tankers, product carriers, drill ships, drilling rigs, and dredgers may be the preferable ship types in this situation.

Figures 6.19–6.21 display a few possible idea solutions for the DC grid. Applying the DC-grid technology might make the cargo pump motors with each VFD easier for a product oil carrier (Figure 6.19). Applying the DC-grid technology might also make the thruster motor section of a pipelayer ship (Figure 6.20) and the winch motor element of a cutter dredger ship (Figure 6.21) easier.

6.7.3 Types of the DC grid

The DC-grid system has been used on a variety of ship types, most notably ferries, tugs, DP ships, and navy ships. Applying the DC-grid system does not require that

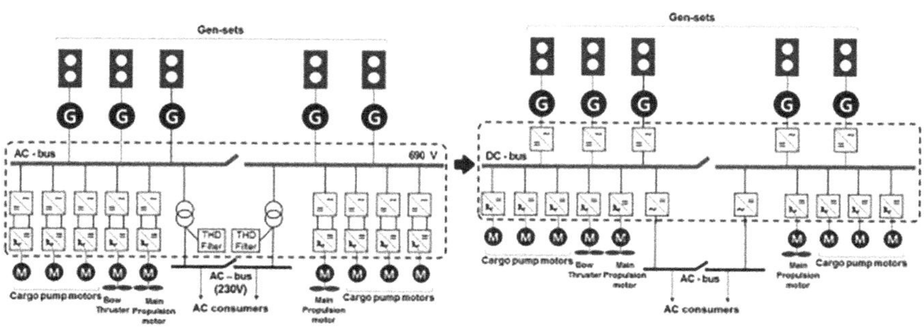

Figure 6.19 Comparison between the conventional AC grid and proposed DC grid of a product oil carrier [3]

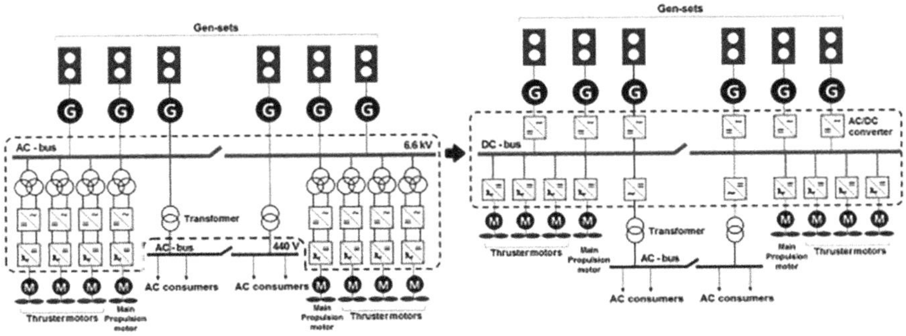

Figure 6.20 Comparison between the conventional AC grid and proposed DC grid of a pipelayer ship [3]

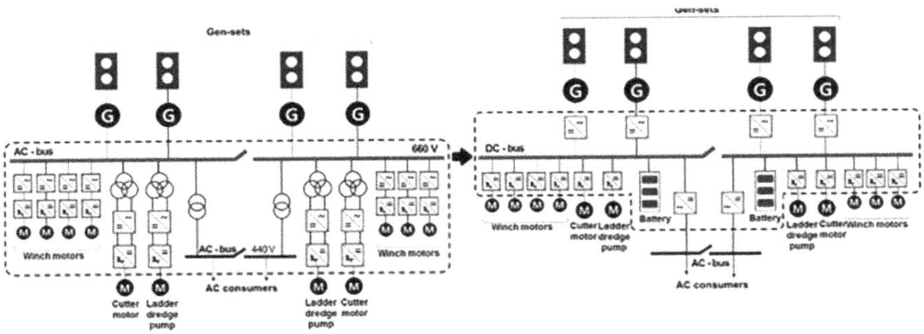

Figure 6.21 Comparison between the conventional AC grid and proposed DC grid of a cutter dredger ship [3]

just the DC be used to distribute electricity throughout the whole system. The DC might be utilized as either a primary or secondary distribution system as:

1. Main-bus type (two-split type, zonal type)
2. Sub-bus type (radial type, common DC-bus type)

The DC can be utilized as a main-bus type first. The most popular and straightforward approach is the two-split type, which may be separated into two categories based on the primary power sources. In the zonal type, a "zone" of one or both of the buses immediately in that compartment supplies power to all loads in that compartment. This guarantees survival since it can stop the spread of harm from fire or flooding. The technology has been used on certain navy ships in conjunction with battery packs or fuel cells as an air-independent propulsion system since it is very redundant. However, this intricate power architecture calls for considerably more advanced control and coordination techniques, which must be carefully considered throughout the early design stages.

Second, the DC has a sub-bus type application. For ships that require to manage several electric motors for certain reasons, the radial type is ideal. For instance, drill strings in a drilling rig must remain steady with respect to the ocean floor. Winch motors raise and lower the string to account for ocean waves in order to maintain proper operation; this process is known as "Active Heave Compensation." This specific component, which consists of several electric motors, can be a DC grid system. Additionally, various special-purpose load components like thruster motors, crane motors, and pump motors might be used.

Between two power converters, there is another form of sub-bus. This kind is typically used on ships that have shaft generator/motors or electric propulsion motors. When the motor is stopped or turned in the opposite direction serving as a generator, it is simple to receive the regenerative electric power if the ESS is attached to this kind (e.g., thrust motors, shaft generators, crane motors, winch motors). In other words, rather than removing the regenerative power through a dynamic brake resistor, the ESS can absorb it.

6.7.4 Technical considerations

Even the DC-grid system has proved to have many benefits over the AC-grid system, it's not common in the marine industry because some of the technical considerations listed below.

(I) Fault protection strategy

When a fault current is interrupted in a DC system, the current must be driven to zero, but in an AC system, the current spontaneously crosses zero twice throughout each period. Due to the sophisticated mechanisms used by DC breakers to properly extinguish the huge fault current and arc, they are more costly and take up more space than AC breakers. The hybrid protection system has been used as a substitute technique in this regard. Fuse, isolating switches, disconnectors, and controlled turn-off of semiconductor power devices are all part of the system's comprehensive protection. It is well known that this type of virtual DC circuit breaker could immediately stop a certain amount of fault current or restrict the amount of fault current that might flow.

However, in crucial locations where strong short-circuit currents may flow and risk a ship's safety, DC circuit breakers should be used for safety and propulsion. Major electric companies have been developing the DC breaking topology in this area to halt a fault's rapid current flow. Previously, the majority of circuit breakers utilized were electromechanical by opening an electromechanical contact and using the comparatively high insulating strength of materials such as air, vacuum, and metal to stop current flow.

The solid-state circuit breaker (SSCB), which uses semiconductor switches instead of electromechanical contacts, has recently been introduced. It can terminate an open circuit when necessary, can transport a lot of

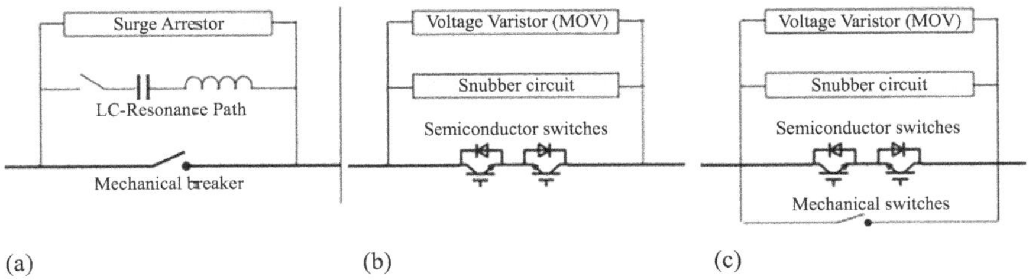

Figure 6.22 Simple schematics of the typical DC circuit breakers [3]. (a) Electro-mechanical circuit breaker, (b) solid-state circuit breaker, and (c) hybrid circuit breaker.

electricity, and switches quickly. Additionally, a hybrid DC circuit breaker has been created that combines the SSCB and the conventional mechanical switch. To operate the semiconductor switches within, however, both the SSCB and the hybrid circuit breaker are necessary, and they are both pricy and challenging at the moment.

Galvanic isolations should also be taken into account to stop any circulating currents in the distribution system. Additionally, grounding is crucial for human safety; thus, the contact voltage should be kept within reasonable bounds. Figure 6.22 shows the different circuit breakers described above.

(II) Power balancing strategy

Both the AC and DC grids depend on the power balancing scheme. In contrast to the AC grid, the DC-grid system does not rely on frequency, and the voltage serves as the sole major signal for active power regulation. Therefore, the PMS should be designed differently from the AC system. Since the voltage angle is related to frequency, frequency can be used to regulate active power. However, in a DC grid, the active power flow is proportional to the DC voltage and is controllable through the DC-link voltage.

In this regard, the frequency–power (*F–P*) droop in the AC grid may be replaced with the voltage–power (*V–P*) or voltage–current (*V–I*) droop scheme as the primary-level power controller in the DC grid, as illustrated in Figure 6.23. In reaction to a change in load, the DC-bus voltage changes, and the droop controller modifies the output power of each converter connected to each power source. To put it another way, this control is accomplished by utilizing both local control of the power electronic interfaces and comprehensive coordination of the complete power system.

Additionally, the droop ratio (droop gain) needs to be chosen in a way that ensures the system's overall stability. When choosing the droop ratio in this regard, the maximum permitted voltage changes are a crucial consideration.

To further enhance power quality, a secondary-level power controller may be employed to make up for any voltage discrepancies brought on by the primary-level controller. When integrating an ESS with bidirectional power

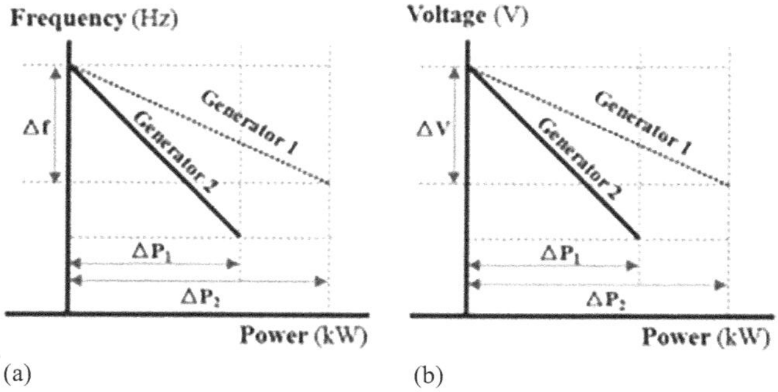

Figure 6.23 Different drooping control strategies. (a) AC-grid system and (b) DC-grid system.

flow (charge or discharge) into a DC grid, it's crucial to take into account the converter's drooping feature because it has two control zones.

(III) Power stability/quality strategy

Since there is no fundamental frequency in a DC-grid system, the harmonic distortion problem that exists in AC-grid systems is not taken into account. However, cyclic or noncyclic load transients may be blamed for most voltage variations that result in power quality problems. The power quality problem may also be brought on by voltage ripple (AC_{rms} over stable DC voltage) from power converters made of controlled switching devices. The IACS has established a standard for voltage quality in Unified Requirement No. E5 (International Association of Classification Societies).

The DC voltage ripple should be kept around 10% since the voltage tolerance is limited to 10%. Additionally, the voltage variation caused by cyclic loads is kept to no more than 5%. The function of converters linked to a DC bus is important in order to achieve these criteria. From the perspective of the entire system, which includes both the AC and DC grids, the harmonics produced by power converters should be kept to the allowable limits. According to IACS, the maximum allowed overall voltage harmonic distortion is less than 8%, and the maximum allowed single-order voltage harmonics varies based on the classification society.

Additionally, the capacity to ride through faults is crucial for preserving power stability. To preserve the healthy units and make sure that the necessary services are provided when one unit in a branch fails, it is important to quickly bypass it. Due to its capacity to respond quickly, the ESS may be a viable alternative to handle an emergency scenario. For speedy responsiveness, it is especially advised to use ESS with high power density.

(IV) Power source control strategy

In order to keep the frequency within the confined working range, a governor must first regulate the speed of the genset in the case of generators in the AC grid utilizing the fixed frequency of 50 or 60 Hz. Nevertheless, in the DC grid, the governor can let the genset to run at various speeds with a predetermined criterion; this criterion may be the minimum amount of fuel consumption (emissions) or the ideal engine loading. In this case, a lockup table with the best speed reference in relation to the engine power might be used to achieve the best speed control and reduce fuel usage.

Additionally, because gensets have their own restricted load acceptance for rapid load disturbances, applying the conventional governor of the AC grid to the DC grid might make the power stability worse. Due to misfiring at lean mixes, or at high air/fuel ratios, the gas engine's capacity for over-boosting is particularly limited. Therefore, unless the ESS is installed to support some amount of power mismatches during the load disturbances, the better transient control mechanism of the governor during dynamic load disturbances should be taken into consideration.

Additionally, there is a good chance that DC-based power sources such as fuel cells, solar cells, lithium-ion batteries, and supercapacitors will be used as the primary or secondary power source in the DC-grid ship. Therefore, it is crucial to understand the characteristics of each power source and power control structure. For instance, the maximum power of solar cells varies with solar radiation, surrounding temperature, and solar cell temperature. As a result, the ESS is integrated with the maximum power point tracking system that controls the majority of its power. Additionally, keeping the overall power stability in mind, the top-level integrated power control system should collaborate with each power control system properly.

(V) DC arc flash hazard

When energy is released as a result of an electrical failure, it might result in an arc flash, which is a hazardous condition. As a result, it is advised to reduce the risk that an arc flash will pose to people in both AC and DC grids. Regarding this, the IEEE standard 1584 offers methods for designers and facility managers to use in figuring out the AC arc flash danger distance and the incident energy to which workers could be exposed when working on or near electrical equipment. However, there are currently no agreed-upon criteria or recommendations for DC arc flashes.

The following technique for calculating DC arc flash incident energy was published in the IEEE. This approach is predicated on the idea that a DC arc would produce its greatest power when the arcing voltage is halved compared to the system voltage. Testing carried out for Bruce Power has demonstrated that this computation is conservatively high in terms of calculating the arc flash magnitude. Up to 1,000 V DC systems can use this technique. In order to prevent HVDC arc flash accidents, it is necessary to study the higher DC arc calculating technique. Additionally, during the design stage, several techniques for detecting, removing, or limiting the DC arc flash must be taken into account.

(VI) Standards or guidelines

The industry must also be given the relevant international standards or guidelines. Despite the fact that the DC-grid system is less popular than the AC grid, there are some available standards and recommendations. For the medium-voltage DC (MVDC)-grid system, revision 1709 was issued. And in May 2017, the IEC 63108 specification was recently made available to the general public. IEC 61660 also covers the process for determining short-circuit currents in a DC system.

Additionally, the US Naval Sea Systems Command (NAVSEA) started creating new MILSTD-1399 sections for low-voltage DC (LVDC) and medium-voltage DC (MVDC) distribution. In order to get input on the early draft of the standards from business, academia, and the government, NAVSEA also published a call for information. For the DC-grid system on ships, classification societies also have some components or an extra preliminary regulation.

6.7.5 Future prospects

The LVDC system is freely accessible on the market, and several references have shown how much more reliable and efficient it is. However, it would be challenging to use the HVDC system for huge ships. The main problem is that high-power converters and HVDC breakers are now costly and not yet generally accessible.

However, it is anticipated that between 2017 and 2021, the worldwide DC system market would expand at a CAGR of 4.98%, with the HVDC industry being particularly promising. Growing demands for RESs as a result of environmental issues and rising electric demand are the main forces behind this increase. If these trends continue, it could further encourage the development of DC-related equipment. Additionally, it would have a favorable impact on the marine industry's HVDC infrastructure.

Although it may be able to deploy the HVDC system in ships in the near future, the loads would be based on traditional AC machines since the DC rotating machine has greater maintenance expenses than AC machines. In order to make the AC and DC systems more effective and secure, they must be coordinated.

6.7.6 Conclusion

The maritime sector has adopted the DC grid and coupled it with the traditional AC grid in response to recent interest in more efficient and dependable power delivery. The advantages of the DC-grid system over the AC grid are discussed in this essay. The flexibility of frequency (reactive power), simplicity of the VFD for motor controls, and ease of integrating DC power sources are the key advantages. And thanks to these advantages, DC-grid ships are now more productive, affordable, and environmentally benign than before. According to type, different ships that have previously implemented or intend to implement the

DC-grid system are addressed. Various DC-grid systems are discussed according to their kind.

The dependability of the DC-grid system in ships is also improved by reviewing a few technical factors. For the DC-grid system in particular, the highly dependable protection and power-sharing topology should be tailored, and the control strategy for each power source should also be taken into account.

Even if the DC-grid system still faces several obstacles before reaching the same standards as the AC-grid system, its future in the maritime sector is promising due to the high need for an eco-friendly or efficient power system in accordance with more stringent environmental rules. All family members, including ship designers, makers, owners, and operators on board, are advised to become familiar with this new approach in the meantime.

6.8 A series-resonant balancing converter for bipolar DC grids on ships

6.8.1 Introduction

One of the primary problems that threaten human life worldwide is climate change. Emissions of GHGs are thought to be the major cause of this. Around 3% of the global GHG emissions are produced by the shipping sector. The IMO has issued regulations requiring ship manufacturers to improve the efficiency of future ships in order to cut emissions. The electrification of the power train is one method for increasing the effectiveness of ships. The main option for this is DC grids on ships. These grids benefit from improved dynamic performance, peak shaving, zero emission operation, and strategic loading, among other things.

Both unipolar and bipolar dc grids are possible. Compared to unipolar dc grids, bipolar dc grids have a number of benefits. Multiple conductors/poles increase dependability because if one fails, the other may be utilized to deliver half of the power. Because there are several voltage levels, they are more flexible. Bipolar DC grids are thus being taken into consideration for future DC grids aboard ships.

Bipolar DC grids have balancing converters as a standard component. In the literature, there are several topologies for these converters. Using a three-level converter to connect the synchronous generator is one of the techniques to create a bipolar DC grid aboard ships. The three-phase multilevel converter topologies like Neutral Point Clamped and T-Type converter can be used to perform the balancing function. These converters may be managed in a way that prevents current from flowing through the neutral and maintains a balanced voltage on the DC side. However, an unbalanced DC operation causes a rise in the overall harmonic distortion on the AC side, which increases the losses in the machine connected to the converter.

Buck-boost, Cuk, and SEPIC converters are further balancing converter topologies. In a bipolar DC grid, these topologies, which are known as inverting topologies, may be used directly as balancing converters. A series-resonant converter topology is an additional fascinating topology. A non-inverting type topology is used in this converter. Therefore, without certain adjustments, it cannot be utilized as a balancing converter. We created a series-resonant balancing converter in this study to demonstrate its usefulness as a balancing converter in bipolar DC grids.

6.8.1.1 Need for balancing converters

For bipolar DC grids, balancing converters are necessary when uneven loads are connected between the +n and n− poles. The voltage at the system neutral shifts as a result of unequal power loading. There are several ways to link the loads to the distribution lines in a bipolar grid with three conductors. The loads may be linked between a pole (+ or −) and neutral or between the positive (+) and negative (−) poles. If there are any imbalanced connections of loads present in an isolated system (IT system), the neutral voltage may deviate from zero. It should be noted that the system lacks a balancing converter. Furthermore, the neutral voltage is not regulated by the droop-controlled converter; only the voltage between the two poles is.

6.8.1.2 Converter operating principle

Figure 6.24 shows the topology used by a converter that is series-resonant. The one shown in the figure is usually used as a noninverting, but for this specific application, it is used as inverting.

The operation of the principle is detailed in Figure 6.25 and the following descriptions.

- $t_0 - t_1$: Switches S1 and S3 are conducting in this condition. The resonant inductor L_r limits current while the resonant capacitor C_r charges.
- $t_1 - t_2$: Switch S3 is off at this time, and switches S1 and S4 are conducting. As a result, the voltage across the resonant tank is greater. The charge is fed into the resonant capacitor as a result of this stage. The capacitor's voltage is increased using the excess charge that has been stored inside.

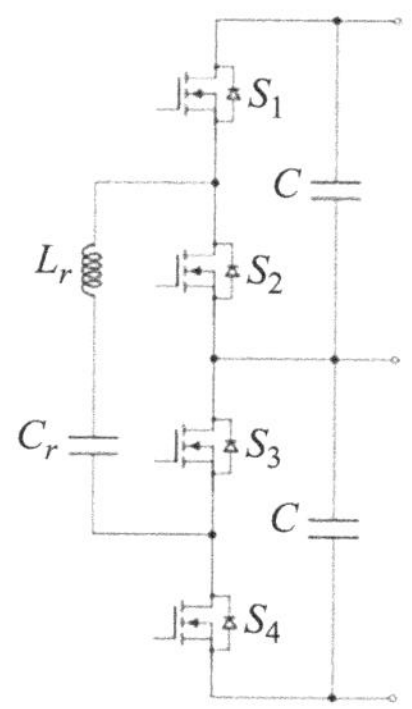

Figure 6.24 Balancing converter based on a series-resonant converter topology [4]

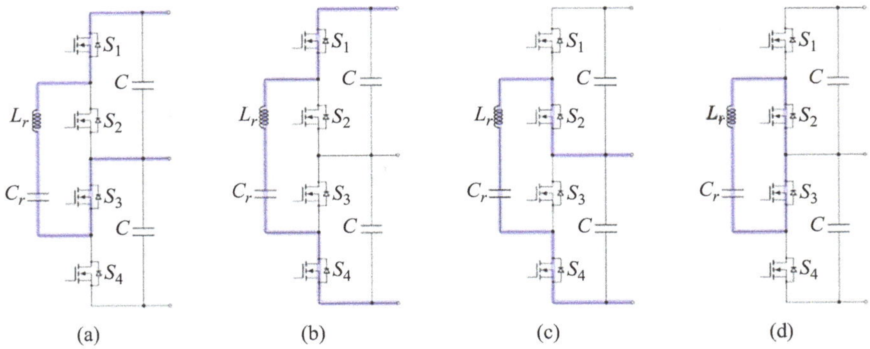

Figure 6.25 Operation of the resonant converter at different time periods [4]. (a) t_0–t_1, (b) t_1–t_2, (c) t_2–t_3, and (d) t_3–t_4.

- $t_2 - t_3$: Switch S1 is off while S2 is on in this situation. Switches S2 and S4 link the resonant tank to the neutral and negative pole. As the energy is transmitted to the load, the inductor current changes direction and the capacitor likewise discharges.
- $t_3 - t_4$: Switch S4 is off while S3 is on in this situation. As a result, the switch diodes are passed through the resonant tank current.

It should be noticed that a phase shift between the top half-bridge (HB) (composed of switches S1 and S2) and lower HB can produce the stages t1–t2 and t3–t4 (consisting switches S3 and S4).

6.8.1.3 Converter prototype

The specifications of the designed prototype are as follows;

- Rated power 2 kW
- Resonant inductance 7 μH
- Resonant capacitance 594 nF

6.8.2 Results

The source converters, load converter, balance converter, and lumped line elements are essential parts. A bipolar DC supply is produced using source converters. In our arrangement, the source converter is an Agilent Technologies N5772A. To mimic an entirely unbalanced power flow, the load converter is merely linked between the neutral and negative poles. When the system is turned on without the balancing converter, just the source converter, which is linked between the neutral and negative poles, is used to supply electricity. The electricity coming from both source converters is balanced using the balancing converter. The current going through the two source converters should be equal when the balancing converter is turned on. The schematic for the balancing converter is shown in Figure 6.26.

LTSpice was used to simulate the converter model together with the other setup components. Figure 6.27 displays the resultant resonant capacitor voltage and resonant inductor current. The load in this simulation is wired between the negative and neutral poles. A current sink with a value of 5 A is the load. The source converter is set to deliver a voltage of about 350 V_{DC}. The switches run at a 50% duty cycle at 65 kHz. Between the switch pair S1, S2, and pair S3, S4, a dead period of 100 ns is added. Phase shift of 3.6° is established between the top HB and bottom HB.

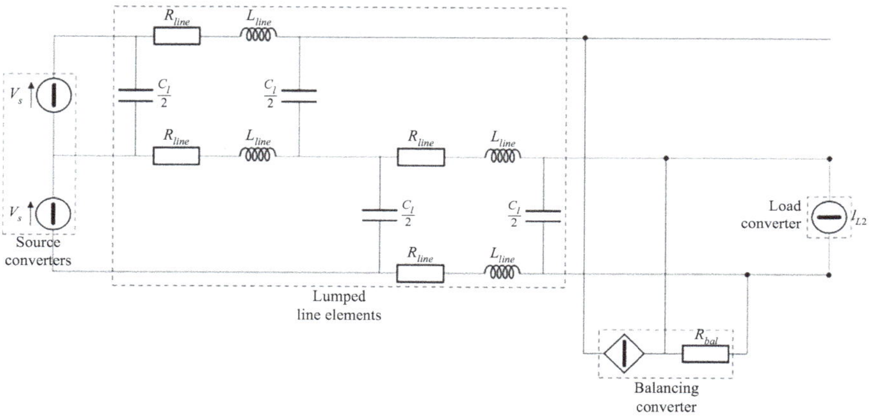

Figure 6.26 Test bench schematic diagram [4]

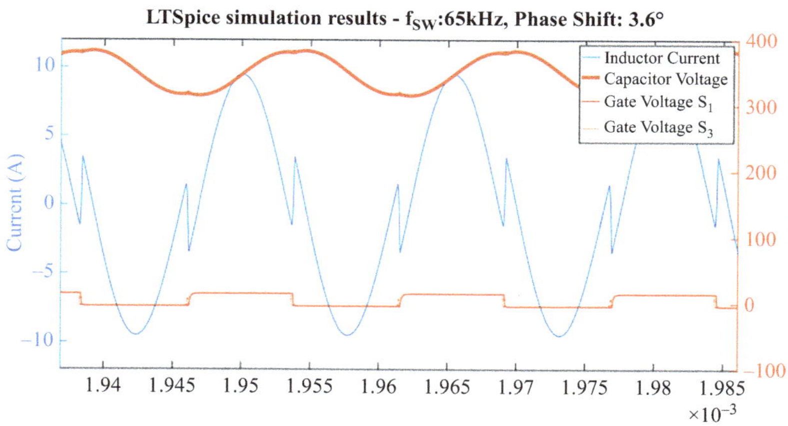

Figure 6.27 Results of the LTSpice simulation [4]

6.8.3 Conclusion

This chapter illustrates how to balance a bipolar DC grid using a series-resonant converter. The suggested converter works in a capacitive area. Power flow may be altered by adjusting the converter's switching frequency and the phase difference between its upper and lower HBs. The balance simulation results for a 1.75 kW load linked between the neutral and negative poles are displayed. To test the simulation results experimentally, a prototype of the balancing converter was constructed. Although not perfectly, the experimental results nearly match the calculations. There is additional discussion about the cause of the discrepancy. The converter will eventually get a closed-loop control designed.

6.9 Modeling and control of the isolated DC–DC modular multilevel converter for electric ship medium-voltage direct current power system

For the contemporary electric ship's medium-voltage dc power system, the DC–DC modular multilevel converter (DC–DC (MMC)) with a medium-frequency (MF) AC-link transformer is a potential choice to provide galvanic isolation, variable voltage control, and fault ride through capability. However, because it has several state variables, the steady states and dynamics of such an isolated DC–DC MMC are rather complex. Additionally, the system design is made more difficult by the on-ship needs. Thus, this work suggests modeling and controlling a similar DC–DC MMC. The fundamental period averaging (FPA) method is used to obtain the steady-state and small-signal models for a DC–DC MMC because it is controlled by phase-shifted operation. Based on the needs of the ship, the modulation index control and simple but effective competitive dual closed loops (a voltage loop for normal condition and a current loop for overload condition) are also designed. It is demonstrated that the suggested controllers may be well developed using the steady-state and small-signal FPA models without sacrificing the on-ship requirements by maintaining the relevant model knowledge about the DC–DC MMC. Simulations and experiments are used to validate the analysis and design outcomes.

6.10 Experimental results of source modeling and simulations

6.10.1 Experimental setup and design procedure

A low-power experimental system is also constructed, as illustrated in Figure 6.28, to further support the study. The steady-state waveforms of the experimental setup are shown in Figure 6.29. The arm voltages and currents that are predominated by the DC and fundamental AC components are shown in Figure 6.29(a). Both the primary and secondary sides' modulation indices – M_{dp} and M_{ds} – are 0.5; as a result, V_{arm_Up} is a staircase voltage with Np equal to four steps and V_{arm_Us} is a

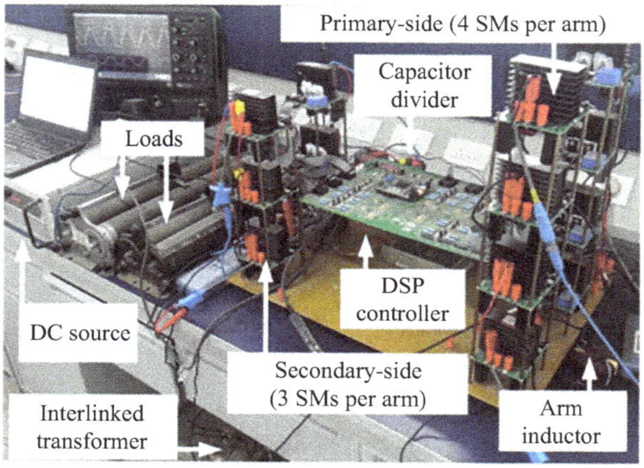

Figure 6.28 Experimental setup [5]

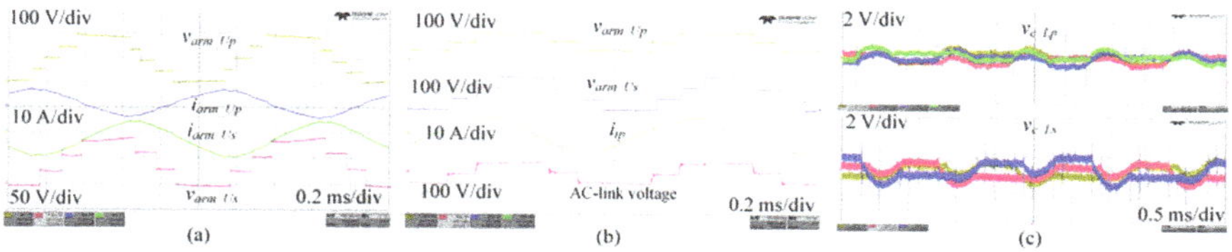

Figure 6.29 Key waveforms in steady state [5]

staircase voltage with Ns equal to three steps. The scenario when M_{dp} is modified to 0.35 and M_{ds} remains at 0.5 is shown in Figure 6.29(b). While V_{arm_Us} remains intact, V_{arm_Up} is modulated as a staircase voltage with two steps to minimize its basic AC component.

In Figure 6.29(b), the voltage and current of the ac-link on the primary side are likewise depicted, and it is clear that they are in ac forms. The SM capacitors in the primary and secondary upper arms are also shown in Figure 6.29(c). They appear to be in good balance. These steady-state waveforms demonstrate the accuracy of the analysis based on our research. The primary design methods and design outcomes are then succinctly described using the analysis that has been given.

- Step 1: The potential operating points and the internal current stresses are analyzed using the circuit characteristics and specifications, steady-state operating point data, and the flowchart in this source.
- Step 2: In this source, the eigenvalues of matrix A are used to survey open-loop stability, and then Formula 25 is used to construct the PI controllers. The experimental case's closed-loop bandwidth is estimated by the Bode diagram analysis to be around 20 rad/s with a minor overshoot.

- Step 3: A modulation index control is added to enhance transient regulation. According to the study, the control method described in (21) of this source also works in the experimental situation.

6.10.2 Open-loop test

The top of Figure 6.30 displays the results of the open-loop test. The open-loop response of V_o is depicted in Figure 6.30(a), where is first set at 0.28 and subsequently stepped to 0.6. It is seen that V_o increases from 85 to 115 V in 0.4 s; the rate of increase is governed by a built-in characteristic of the DC–DC MMC. The transfer function's response with the same step is shown at the bottom, and it can be observed that the amplitude step and increasing rate are consistent with the results of the studies. In Figure 6.30, comparisons of additional important waveforms are also provided, and they are similarly very well matched.

6.10.3 Closed-loop test

On the top of Figure 6.31, the voltage closed-loop experimental waveforms are displayed. The voltage reference V_o in Figure 6.31(a) ranges from 90 to 100 V. V_o can track its reference within a tiny overshoot thanks to the closed-loop effect. The

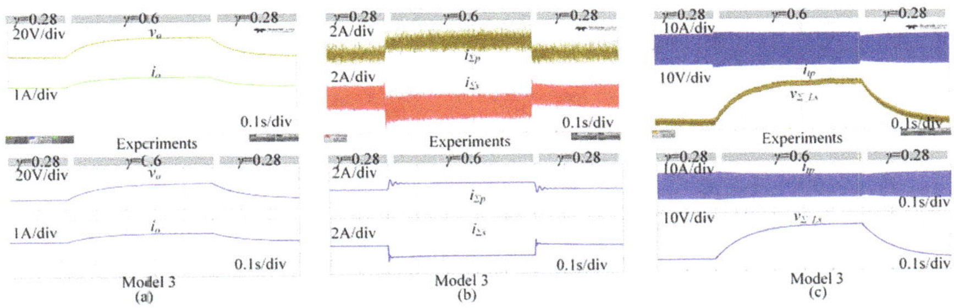

Figure 6.30 Comparison of the open-loop performances [5]

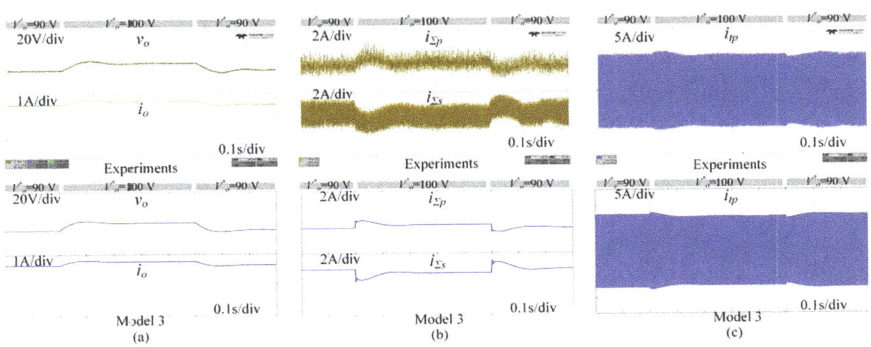

Figure 6.31 Comparison of the voltage closed-loop performances [5]

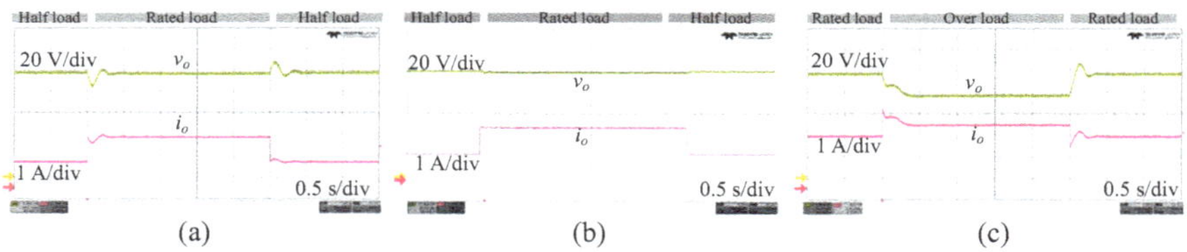

Figure 6.32 Dynamical performances under load variation [5]

Model 3 replies, which are likewise displayed at the bottom, closely reflect the outcomes of the experiment. In Figure 6.31, comparisons of additional important waveforms are also provided. These comparisons show that they are highly matched, supporting the validity of the models and the controller design process.

Figure 6.32 displays the control performances under changing loads. The output is operating at half-load in Figure 6.32(a), and V_o tracks to its reference V_o. Once the load has been increased to the rated load, the voltage closed loop will allow V_o to be regulated back to V_o. Because M_{dp} is set at 0.5 in this situation, V_o has a significant overrun. The modulation index control (21) is used in Figure 6.32(b). M_{dp} is managed as 0.35 in the half-load condition and as 0.5 in the full-load situation. As can be observed, there is a significant decrease in V_o overshoot during load changeover.

6.11 Conclusion

The modeling and control problems of the isolated DC–DC MMC are the main topics of this research. An innovative nine-order dynamic model describes the isolated DC–DC MMC on the basis that it has been properly modulated. The FPA technique is then given, in which the state variables in the original model are described in Fourier series form and only the dominant components are kept, based on the observation that the DC–DC MMC is controlled by the phase-shifted angle. The original dynamic model is transformed into a 15-order model using the FPA approach, which considers the basic AC and DC components as new state variables.

The steady-state working point analysis and small-signal analysis are made feasible by the FPA model. The modulation index control is offered based on steady-state analysis by sacrificing the inner current stress and output voltage overshoot. The frequency-domain transfer functions are developed to direct the design of the controllers based on the small-signal analysis. The twin closed loops are highly suited to respond to the requirements of the on-ship applications because of the FPA concept.

The isolated DC–DC MMC's switching, original dynamic, and FPA models are all constructed in the simulation for comparison, and a scaled-down prototype is constructed in the lab to further confirm the correctness. The results from the simulation and the experiment are in good agreement with the analysis. The isolated DC–DC MMC can provide adequate voltage regulation in the normal load situation and restrict the current in the overload condition with the chosen modulation index plus phase-shifted control mechanism.

References

[1] Jin, Z., Sulligoi, G., Cuzner, R., *et al.* (2016). "Next-generation shipboard DC power system: Introduction smart grid and DC microgrid technologies into maritime electrical networks." *IEEE Electrification Magazine*, 4(2), 45–57. https://doi.org/10.1109/mele.2016.2544203.

[2] Sulligoi, G., Bosich, D., Giadrossi, G., Zhu, L., Cupelli, M., and Monti, A. (2014). "Multiconverter medium voltage DC power systems on ships: Constant-power loads instability solution using linearization via state feedback control." *IEEE Transactions on Smart Grid*, 5(5), 2543–2552. https://doi.org/10.1109/tsg.2014.2305904

[3] Kim, K., Park, K., Roh, G., and Chun, K. (2018). "DC-grid system for ships: A study of benefits and technical considerations." *Journal of International Maritime Safety, Environmental Affairs, and Shipping*, 2(1), 1–12. https://doi.org/10.1080/25725084.2018.1490239

[4] Yadav, S., Qin, Z., and Bauer, P. (1970, January 1). "A series resonant balancing converter for bipolar DC grids on ships." TU Delft Research Portal. Retrieved October 21, 2022, from https://research.tudelft.nl/en/publications/a-series-resonant-balancing-converter-for-bipolar-dc-grids-on-shi.

[5] Chen, Y., Zhao, S., Li, Z., Wei, X., and Kang, Y. (2017). "Modeling and control of the isolated DC–DC modular multilevel converter for electric ship medium voltage direct current power system." *IEEE Journal of Emerging and Selected Topics in Power Electronics*, 5(1), 124–139.

Chapter 7

Regulation and standardization of the electric ships

Mohamed Ebeed[1,2], *Francisco Jurado*[2] *and Shady H.E. Abdel Aleem*[3]

7.1 Electric ship market

With increasing global commercial trading, the electric ship market is increasing rapidly. By 2030, the expected increase of the electric ship market is 14.2 billion USD or 20% compared to 2022 which is 3.3 billion USD. Figure 7.1 illustrates the increase of the electric ship market to 2026 [1,2]. The main key drivers of the electric ship market are increasing fossil fuel prices, world trends to reduce the emissions from transportation systems and maximizing the dependence on renewable energy sources (RES), as well as the progress of the new technologies of the batteries, fuel cell, and the RES.

It should be highlighted here that the segmentation of the electric ship market depends upon the ship types including passenger or commercial ships. The operation mode of the electric ships includes the semi-autonomous or fully autonomous mode. Furthermore, the propulsion types of the electric ship include hybrid or fully electric.

7.2 Marine electrical standard definitions

Vessels need to operate safely and efficiently, especially the electric system in ships. Thus, the boats and ship owners should follow the guidelines to avoid accidents and maintain the integrity of maritime operations. In this regard, the marine electrical standards represent the guidelines that should be followed by the owners or the manufacturers.

[1]Electrical Engineering Department, Faculty of Engineering, Sohag University, Egypt
[2]Department of Electrical Engineering, University of Jaen, Spain
[3]Department of Electrical Engineering, Institute of Aviation Engineering and Technology, Egypt

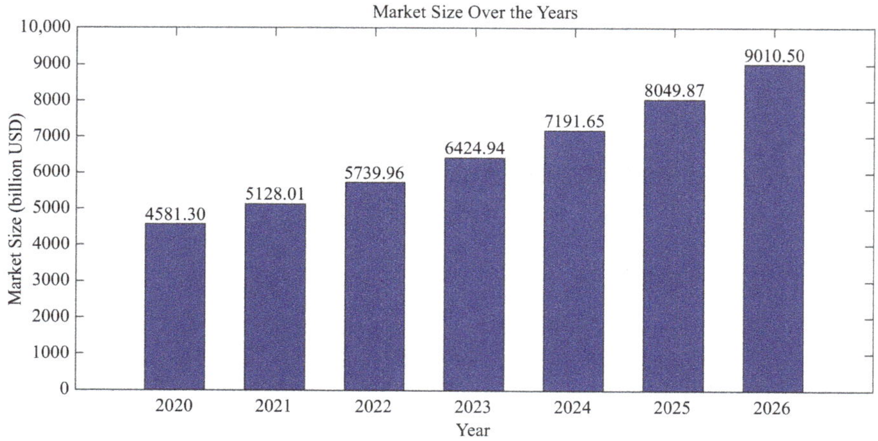

Figure 7.1 The annual increase in the market size of the electric ships over the year

7.2.1 International Electrotechnical Commission (IEC)

The International Electrotechnical Commission (IEC) is an international organization, and it sets the standards of electronic, electrical, as well as related technologies. In addition to that, the IEC can set and develop the standards of electrical installation in boats and ships, including switchgear, circuit breakers, wiring, control circuits, and navigation equipment. The important benefits of applying the IEC include ensuring safety where the suitable installation and design can reduce electrical hazards. Also, it can optimize the operation of vessels by reducing fuel consumption and emissions. In addition to that, following the marine electrical standards can encourage global maritime commerce between countries where the manufacturers of the vessels follow the same standards.

The International Telecommunication Union (ITU) and The International Organization for Standardization (ISO) are similar organizations that also set and develop International Standards in other industries. The three organizations (IEC, ISO, and ITU) coordinate when their standardization activities overlap.

7.2.2 The National Electric Code (NEC)

The National Electric Code (NEC) is a set of standards for equipment, installation, design, grounding, and electric wiring installation in the United States and the NEC is known as NFPA 70 (the National Fire Protection Association) [3,4]. This code is updated every three years.

7.2.3 The American Boat and Yacht Council (ABYC) safety

The American Boat and Yacht Council (ABYC) is a non-profit organization that sets and develops voluntary safety standards for the maintenance, construction, and repair the recreational boats and ships [5]. These standards include the boats' design, electrical systems, plumbing, fuel systems, and ventilation to ensure the reliability and safety of the boats. The ABYC standards are important for marine technicians, manufacturers, and industry professionals. The ABYC updates its standard periodically for boating safety. The ABYC Certification is very important for boat manufacturers to guarantee that they follow safety standards.

7.2.4 The International Maritime Organization (IMO)

The IMO is an agency that was established in 1948 and 1958; it became a specialized agency of the United Nations [6]. The aims and roles of the IMO agency can be outlined as follows:

- The IMO regulates shipping on a global scale and encourages safe, secure, efficient, and eco-friendly shipping and navigation.
- The IMO is responsible for facilitating international maritime traffic.
- The IMO amends and adopts the protocols and the international conventions and sets standards for the security safety, and environmental performance of international navigation and shipping which represents the guidelines for all the member states.
- The IMO consists of technical Committees and sub-committees which are responsible for reviewing guidelines, recommendations, and regulations and focuses on specific aspects of shipping. The regulations key to the IMO include the Safety of Life at Sea (SOLAS), the International Safety Management Code (ISM Code), and the International Convention for the Prevention of Pollution from Ships (MARPOL).
- The IMO presents technical assistance for developing shipping standards in different countries.

7.3 IEC technical committees (TCs)

It should be highlighted here that two technical committees are responsible for specifying the standard in the marine sector including the IEC TC 80 [7] which is responsible for setting and developing the standards of radio communications equipment and maritime navigation systems. Furthermore, this TC is responsible for putting the IEC 61097 *Global Maritime Distress and Safety System* (GMDSS) series of standards relating to ship safety systems for different components. The second committee is the TC 18 [8] which is responsible for electrical installations standards of ships of mobile and fixed offshore systems; also this TC prepared the IEC 60092 series which accommodates the International Convention on Safety of Life at Sea (SOLAS).

Also, there are many IEC TCs that are responsible for developing the IEC standard for the shipping sector [9]. Table 7.1 summarizes the IEC standards codes that should be applied to the marine sector.

Table 7.1 The working IEC standard codes for the marine sector

The TC	Responsibilities	The corresponding standards and codes
IEC TC 80 [7]	Maritime navigation and radiocommunication equipment and systems.	• **IEC 61097-4 ED4:** Global maritime distress and safety system (GMDSS) – Part 4: Inmarsat-C ship earth station and Inmarsat enhanced group call (EGC) equipment – Operational and performance requirements, methods of testing and required test results. • **IEC 61097-7 ED2:** Global maritime distress and safety system (GMDSS) – Part 7: Shipborne VHF radiotelephone transmitter and receiver – Operational and performance requirements, methods of testing, and required test results. • **IEC 61097-9 ED2:** Global maritime distress and safety system (GMDSS) – Part 9: Shipborne transmitters and receivers for use in the MF and HF bands suitable for telephony, digital selective calling (DSC) and reception of Maritime Safety Information and Search and Rescue related information – Operational and performance requirements, methods of testing and required test results. • **IEC 61108-7 ED1:** Maritime navigation and radiocommunication equipment and systems – Global navigation satellite systems (GNSS) – Part 7: Satellite-based augmentation system (SBAS) L1 – Receiver equipment – Performance standards, methods of testing and required test results. • **IEC 61162-1 ED6:** Maritime navigation and radiocommunication equipment and systems – Digital interfaces – Part 1: Single talker and multiple listeners • **IEC 61162-2 ED2:** Maritime navigation and radiocommunication equipment and systems – Digital interfaces – Part 2: Single talker and multiple listeners, high-speed transmission. • **IEC 61162-450 ED3**: Maritime navigation and radiocommunication equipment and systems – Digital interfaces – Part 450: Multiple talkers and multiple listeners – Ethernet interconnection.

(Continues)

		• **IEC 61162-460 ED3:** Maritime navigation and radiocommunication equipment and systems – Digital interfaces – Part 460: Multiple talkers and multiple listeners – Ethernet interconnection – Safety and security. • **IEC 61174 ED5:** Maritime navigation and radiocommunication equipment and systems – Electronic chart display and information system (ECDIS) – Operational and performance requirements, methods of testing, and required test results. • **IEC 62065 ED3:** Maritime navigation and radiocommunication equipment and systems – Track control systems – Operational and performance requirements, methods of testing, and required test results. • **IEC 63514 ED1:** Maritime navigation and radiocommunication equipment and systems – VHF Data Exchange System (VDES) – Shipborne mobile station – Operational and performance requirements, methods of test, and required test results.
IEC TC 18 [8]	Electrical installations of ships and mobile and fixed offshore units.	• **PNW 18-1816 ED1:** Utility connections in port – Part 4: DC shore connection (DCSC) systems – General requirements • **IEC 60092-302-2 ED2:** Electrical installations in ships – Part 302-2: Low voltage switchgear and control gear assemblies – Marine Power • **IEC 60092-305 ED4:** Electrical installations in ships – Part 305: Equipment – Accumulator (storage) batteries • **IEC 60092-401 ED4:** Electrical installations in ships – Part 401: Installation and test of completed installation. • **IEC 60092-501 ED6:** Electrical installations in ships – Part 501: Special features – Electric propulsion plant. • **IEC 60092-502 ED6:** Electrical installations in ships – Part 502: Tankers – Hazardous area classification and special features. • **IEC 60092-504 ED6:** Electrical installations in ships – Part 504: Automation, control and instrumentation. • **IEC 60092-507 ED4:** Electrical installations in ships – Part 507: Small vessels.

(Continues)

Table 7.1 (Continued)

The TC	Responsibilities	The corresponding standards and codes
		• **IEC 60092-509 ED2:** Electrical installations in ships – Part 509: Operation of electrical installations. • **IEC 60533 ED4:** Electrical and electronic installations in ships – Electromagnetic compatibility (EMC). • **IEC 61363-1 ED2:** Electrical installations of ships and mobile and fixed offshore units – Part 1: Procedures for calculating short-circuit currents in three-phase AC • IEC/IEEE 61886-2 ED1: Subsea equipment – Part 2: Power transformers. • **IEC 61892-1 ED5:** Mobile and fixed offshore units – Electrical installations – Part 1: General requirements and conditions. • **IEC 61892-2 ED4:** Mobile and fixed offshore units – Electrical installations – Part 2: System design. • **IEC 61892-3 ED5:** Mobile and fixed offshore units – Electrical installations – Part 3: Equipment. • **IEC 61892-4 ED3:** Mobile and fixed offshore units – Electrical installations – Part 4: Cables. • Mobile and fixed offshore units – Electrical installations – Part 5: Mobile units. • **IEC 61892-6 ED5:** Mobile and fixed offshore units – Electrical installations – Part 6: Installation. • **IEC 61892-7 ED5:** Mobile and fixed offshore units – Electrical installations – Part 7: Hazardous areas. • **IEC 63108 ED1:** Electrical installations in ships – Primary DC distribution, system design architecture. • **IEC TR 63436 ED1:** Insulation tests and insulation resistance threshold. • **IEC 63462-1 ED1:** Maritime battery system – Part 1: Secondary lithium cells and batteries – Safety requirements. • **IEC/IEEE 80005-3 ED1:** Utility connections in port – Part 3: low Voltage Shore Connection (LVSC) Systems – General requirements. • **ISO 16315 ED2:** Small craft – Electric propulsion system.

(Continues)

IEC TC 2 [10]	Rotating machinery.	• **IEC 60204-1:** Safety of machinery. Electrical equipment of machines – Part 1: General requirements • **IEC 60404 series:** Magnetic materials. • **IEC 60072 series:** Dimensions of rotating electrical machines. • **IEC 60034 series**: Rotating electrical machines. • **IEC 60079-7**: Explosive atmospheres – Part 7: Equipment protection by increased safety. • **PNW 2-2156 ED1:** Specific test methods for determining losses and efficiency of water submersible motors. • **PNW TS 2-2169 ED1:** Technical Specifications for adjustable speed pumped storage system and its generator motor. • **IEC 60034-1 ED15**: Rotating electrical machines – Part 1: Rating and performance • **IEC 60034-2-1 ED3:** Rotating electrical machines – Part 2-1: Standard methods for determining losses and efficiency from tests (excluding machines for traction vehicles). • **IEC 60034-2-2 ED2:** Rotating electrical machines – Part 2-2: Specific methods for determining separate losses of large machines from tests – Supplement to IEC 60034-2-1. • **IEC 60034-2-3 ED2:** Rotating electrical machines – Part 2-3: Specific test methods for determining losses and efficiency of converter-fed AC motors. • **IEC 60034-12 ED4:** Rotating electrical machines – Part 12: Starting performance of single-speed three-phase cage induction motors. • **IEC 60034-15 ED4:** Rotating electrical machines – Part 15: Impulse voltage withstand levels of form-wound stator coils for rotating AC machines. • **IEC 60034-18-31 ED3:** Rotating electrical machines – Part 18-31: Functional evaluation of insulation systems – Test procedures for form-wound windings – Thermal evaluation and classification of insulation systems used in rotating machines. • **IEC 60034-18-32/AMD1 ED2**: Amendment 1 – Rotating electrical machines – Part 18-32: Functional evaluation of insulation systems (Type II) – Electrical endurance qualification procedures for form-wound windings.

(Continues)

Table 7.1 (Continued)

The TC	Responsibilities	The corresponding standards and codes
		• **IEC 60034-18-41 ED2:** Rotating electrical machines – Part 18-41: Partial discharge free electrical insulation systems (Type I) used in rotating electrical machines fed from voltage converters – Qualification and quality control tests. • **IEC TS 60034-27-6 ED1:** Rotating electrical machines – Part 27-6: On-line partial discharge measurements of rotating machine windings supplied from an inverter. • **IEC 60034-27-7 ED1:** Rotating electrical machines – Part 27-7: Insulation systems used in rotating electrical machines for sealed and moisture-resistant winding type and quality control tests. • **IEC 60034-27-8 ED1:** Rotating machinery – Part 27-8: Detection of inter-turn short-circuits in rotor windings of cylindrical rotor synchronous generator. • **IEC 60034-30-3 ED1:** Rotating electrical machines – Part 30-3: Efficiency classes of high voltage AC motors (IE-code). • **IEC 60034-35 ED1:** Rotating electrical machines – Part 35: Technical requirements for electrical sheet metal and strip metal used in electrical machines. • **IEC TS 60034-38 ED1:** Guidelines for condition-based evaluation and maintenance of cylindrical-rotor synchronous generators. • **IEC 60072-2 ED2**: Dimensions and output series for rotating electrical machines – Part 2: Frame numbers 355 to 1000 and flange numbers 1180 to 2360. • **IEC 60136 ED3:** Dimensions, marking, and testing of carbon brushes and dimensions of brush-holders for electrical machinery. • **IEC 60413 ED2**: Test procedures for determining physical properties of brush materials for electrical machines.

(Continues)

IEC TC 105 [11].	Fuel cell technologies.	• **PWI 105-1:** General safety standard • **PNW 105-1005 ED1:** Fuel cell technologies – Part 4-401: Fuel cell power systems for propulsion and auxiliary power units – Maritime sector – Safety of PEMFC-Systems • **IEC 62282-2-400 ED1:** Fuel cell technologies – Part 2-400: Fuel cell modules – Calculation of Rated Power and Power Density of a PEM stack and PEM module • Fuel cell technologies – Part 3-100: Stationary fuel cell power systems – Safety • Fuel cell technologies – Part 3-200: Stationary fuel cell power systems – Performance test methods • IEC 62282-3-201 ED3: Fuel cell technologies – Part 3-201: Stationary fuel cell power systems – Performance test methods for small fuel cell power systems • **IEC 62282-3-202 ED1:** Fuel cell technologies – Part 3-202: Stationary fuel cell power systems – performance test methods for small fuel cell power systems that can be complemented with a supplementary heat generator for multiple units operation by an energy management system • **IEC 62282-6-101 ED1:** Fuel cell technologies – Part 6-101: Micro fuel cell power systems – Safety – General requirements • IEC 62282-6-106 ED1: Fuel cell technologies – Part 6-106: Micro fuel cell power systems – Safety – Indirect Class 8 (corrosive) compounds • IEC 62282-6-107 ED1: fuel cell technologies – Part 6-107: micro fuel cell power systems – safety – indirect water-reactive (Division 4.3) compounds • **IEC 62282-6-401 ED1:** Fuel cell technologies – Part 6-401: Micro fuel cell power systems – Power and data interchangeability – Performance test methods for laptop computers • **IEC TS 62282-7-1 ED3:** Fuel cell technologies – Part 7-1: Test methods – Single cell performance tests for polymer electrolyte fuel cells (PEFC) • **IEC 62282-7-2 ED2:** Fuel cell technologies – Part 7-2: Test methods – Single cell and stack performance tests for solid oxide fuel cells (SOFCs).

(Continues)

Table 7.1 (*Continued*)

The TC	Responsibilities	The corresponding standards and codes
		• **IEC 62282-8-201 ED2:** Fuel cell technologies – Part 8-201: Energy storage systems using fuel cell modules in reverse mode – Test procedures for the performance of power-to-power systems. • **IEC 63341-3 ED1:** Railway applications – Rolling stock – Part 3: Fuel cell systems for propulsion – Performance test methods.
IEC TC 40 [12]	Capacitors and resistors for electronic equipment.	• IEC 60115-8-1 ED3: Fixed surface mount (SMD) low-power film resistors for general electronic equipment, • IEC 60115-9 ED2: Fixed surface mount resistor networks with individually measurable resistors • IEC 60384-8 ED5: Fixed capacitors for use in electronic equipment – Part 8: Sectional specification: Fixed capacitors of ceramic dielectric, • IEC 60384-9 ED5: Fixed capacitors of ceramic dielectric, • IEC 60384-14-1 ED4: Fixed capacitors for electromagnetic interference suppression and connection to the supply mains • IEC 60384-14-2 ED3: Fixed capacitors for electromagnetic interference suppression and connection to the supply mains – Safety tests only • IEC 60384-21 ED4: Fixed surface mount multilayer capacitors of ceramic dielectric, • surface mount multilayer capacitors of ceramic dielectric. • IEC 60939-3 ED2: Passive filter units for which safety tests are appropriate. • IEC 60940 ED3: Guidance information on the application of capacitors, resistors, inductors, and complete filter units for electromagnetic interference suppression • Sectional specification for surge suppression varistors • IEC 62391-2 ED2: Electric double-layer capacitors for power application • IEC 62813 ED2: Lithium-ion capacitors for use in electric and electronic equipment • IEC TS 63337 ED1: Basic qualification of DC-link film capacitors for automotive.

(Continues)

IEC TC 23 [13]	Electrical accessories.	• **IEC 60364 series**: Electrical installations of buildings. • **IEC 60269 series**: Low-voltage fuses. • **IEC 61058 series**: Switches for appliances. • **IEC 60947 series**: Low-voltage switchgear and control gear. • **IEC 61058 series**: Switches for appliances. • **IEC 60417 series**: Graphical symbols for use on equipment.
IEC TC 82 [14]	Solar photovoltaic energy systems.	• PNW 82-2194 ED1: Building integrated photovoltaic (BIPV) – Identification code for building-integrated photovoltaic modules • PNW 82-2195 ED1: Supplemental Test requirement of building-integrated photovoltaic (BIPV) module containing an additional glass to a certified PV module. • PNW TS 82-2202 ED1: Classification of n-type crystalline silicon photovoltaic cells with EL detected black arc • PNW TS 82-2208 ED1: Electrical safety of Snow melting photovoltaic (Snow PV) module – Requirements for construction and testing. • PNW TS 82-2212 ED1: Coupled-Stress Acceleration Test Sequence for Photovoltaic Modules and Materials • IEC TS 60904-1-2 ED2: Photovoltaic devices – Part 1-2: Measurement of current-voltage characteristics of bifacial photovoltaic (PV) devices • IEC 60904-8 ED4: Photovoltaic devices – Part 8: Measurement of spectral responsivity of a photovoltaic (PV) device • IEC 61215-1/AMD1 ED2: Amendment 1 – Terrestrial photovoltaic (PV) modules – Design qualification and type approval – Part 1: Test requirements • IEC 61215-1-1/AMD1 ED2: Amendment 1 – Terrestrial photovoltaic (PV) modules – Design qualification and type approval – Part 1-1: Special requirements for testing of crystalline silicon photovoltaic (PV) modules

(Continues)

Table 7.1 (Continued)

The TC	Responsibilities	The corresponding standards and codes
		• IEC 61215-2/AMD1 ED2: Amendment 1 – Terrestrial photovoltaic (PV) modules – Design qualification and type approval – Part 2: Test procedures • IEC 61683 ED2: Photovoltaic systems – Power conditioners – Procedure for measuring efficiency. • IEC TS 61724-2 ED2: Photovoltaic system performance – Part 2: Capacity evaluation method • IEC TS 61724-3 ED2: Photovoltaic system performance – Part 3: Energy evaluation method • IEC TS 61836-1 ED1: Solar photovoltaic energy systems – Terms, definitions, and symbols – Part 1: Basic terms • IEC 61853-1/AMD1 ED1: Amendment 1 – Photovoltaic (PV) module performance testing and energy rating – Part 1: Irradiance and temperature performance measurements and power rating • IEC 61853-2 ED2: Photovoltaic (PV) module performance testing and energy rating – Part 2: Spectral responsivity, incidence angle, and module operating temperature measurements • IEC 61853-3/AMD1 ED1: Amendment 1 – Photovoltaic (PV) module performance testing and energy rating – Part 3: Energy rating of PV modules • IEC 61853-4/AMD1 ED1: Amendment 1 – Photovoltaic (PV) module performance testing and energy rating – Part 4: Standard reference climatic profiles. • IEC 62109-1 ED2: Safety of power converters for use in photovoltaic power systems – Part 1: General requirements. • IEC 62109-2 ED2: Safety of power converters for use in photovoltaic power systems – Part 2: Particular requirements for inverters. • IEC 62116 ED3: Utility-interconnected photovoltaic inverters – Test procedure of islanding prevention measures • IEC 62253 ED2: Photovoltaic pumping systems – Design qualification and performance measurements

(Continues)

- IEC TS 62257-9-5 ED5: Renewable energy and hybrid systems for rural electrification – Part 9-5: Integrated systems – Laboratory evaluation of stand-alone renewable energy products for rural electrification
- IEC TS 62257-9-8 ED2: Renewable energy and hybrid systems for rural electrification – Part 9-8: Integrated systems – Requirements for stand-alone renewable energy products with power ratings less than or equal to 350 W
- IEC TS 62257-301 ED1: Renewable energy off-grid systems – Part 301: Generators – Integration of solar with other forms of power generation within hybrid power systems
- IEC TS 62257-341 ED1: Renewable energy off-grid systems – Part 341: Selection of batteries and battery management systems for stand-alone electrification systems – Specific case of automotive flooded lead-acid batteries available in developing countries
- IEC TS 62257-350 ED1: Renewable energy off-grid systems – Part 350: Recommendations for selection of inverters
- Photovoltaic (PV) systems – Requirements for testing, documentation, and maintenance – Part 1: Grid-connected systems – Documentation, commissioning tests, and inspection
- IEC TS 62446-3 ED2: Photovoltaic (PV) systems – Requirements for testing, documentation, and maintenance – Part 3: Photovoltaic modules and plants – Outdoor infrared thermography
- IEC TS 62446-4 ED1: Photovoltaic (PV) systems – Requirements for testing, documentation, and maintenance – Part 4: Electroluminescence Measurement of Photovoltaic Arrays
- IEC 62509 ED2: Battery charge controllers for photovoltaic systems – Performance and functioning
- IEC 62688 ED2: Concentrator photovoltaic (CPV) modules and assemblies – Safety qualification
- IEC 62788-1-1 ED1: Measurement procedures for materials used in photovoltaic modules – Part 1-1: Encapsulants – Polymeric materials used for encapsulants
- IEC TS 62788-2 ED2: Measurement procedures for materials used in photovoltaic modules – Part 2: Polymeric materials – Frontsheets and back sheets

(Continues)

Table 7.1 (Continued)

The TC	Responsibilities	The corresponding standards and codes
		• IEC TS 62788-7-2/AMD1 ED1: Amendment 1 –- Measurement procedures for materials used in photovoltaic modules – Part 7-2: Environmental exposures – Accelerated weathering tests of polymeric materials • IEC 62788-7-3/AMD1 ED1: Amendment 1 – Measurement procedures for materials used in photovoltaic modules – Part 7-3: Accelerated stress tests – Methods of abrasion of PV module external surfaces • IEC TS 62788-8-1 ED1: Measurement procedures for electrically conductive adhesive (ECA) used in crystalline silicon photovoltaic modules – Part 8-1: Measurement of material properties • IEC TS 62804-1 ED2: Photovoltaic (PV) modules – Test methods for the detection of potential-induced degradation – Part 1: Crystalline silicon • IEC 62817-1 ED1: Solar photovoltaic tracking systems – Part 1: Design qualification for horizontal one-axis solar tracking system • IEC 62852 ED2: Connectors for DC-application in photovoltaic systems – Safety requirements and tests • IEC 62894 ED2: Photovoltaic inverters – Datasheet and nameplate • IEC 62920 ED2: Photovoltaic power generating systems – EMC requirements and test methods for power conversion equipment • IEC 62941 ED2: Terrestrial photovoltaic (PV) modules – Quality system for PV module manufacturing • IEC 63092-3 ED1: Photovoltaics in buildings – Part 3: Evaluation methodology of SHGC for Building integrated photovoltaic modules with various designs • IEC TS 63126 ED2: Guidelines for qualifying PV modules, components, and materials for operation at high temperatures • IEC TS 63202-6 ED1: Photovoltaic cells – Part 6: Hot water soaking test for crystalline silicon solar cells • IEC 63257 ED1: Power line communication for DC shutdown equipment – Communication signal, physical layer

(Continues)

- IEC TS 63348 ED1: Evaluation of Photovoltaic (PV) Module to Mounting Structure Interface
- IEC 63349-1 ED1: Photovoltaic direct-driven appliance controllers – Part 1: General Requirement
- IEC TS 63371-1 ED1: Materials used in photovoltaic (PV) cells – Part 1: Specifications for electrical characteristics of crystalline silicon wafers.
- IEC 63387-1 ED1: Hybrid CPV/PV modules: General characteristics and measurement procedures – Part 1: Performance measurements and power rating – Irradiance and temperature
- Fire test for concentrator PV modules
- IEC 63409-1 ED1: Photovoltaic power generating systems connection with grid – Conformity assessment for power conversion equipment – Part 1: Overall description of conformity assessment for grid connection
- IEC 63409-2 ED1: Photovoltaic power generating systems connection with grid – Testing of power conversion equipment – Part 2: Testing environment
- IEC 63409-3 ED1: Photovoltaic power generating systems connection with grid – Testing of power conversion equipment – Part 3: Basic operations
- IEC 63409-4 ED1: Photovoltaic power generating systems connection with grid – Conformity assessment for power conversion equipment – Part 4: Interface protection and fault ride through.
- IEC 63409-5 ED1: Photovoltaic power generating systems connection with grid – Testing of power conversion equipment – Part 5: Power Quality and EMC
- IEC 63409-6 ED1: Photovoltaic power generating systems connection with grid – Conformity assessment for power conversion equipment – Part 6: Power control functions and grid support.
- IEC 63409-7 ED1: Photovoltaic power generating systems connection with grid – Testing of power conversion equipment – Part 7: Information exchange
- IEC TS 63451 ED1: Power conditioners efficiency with partially shaded photovoltaic generators
- IEC TS 63496 ED1: Floating photovoltaic power plants – Design guidelines and recommendations
- Solar Trackers – Requirements for the protection of personnel
- Reuse of PV modules and circular economy

(Continues)

Table 7.1 (*Continued*)

The TC	Responsibilities	The corresponding standards and codes
		• IEC TS 63543-1 ED1: Photovoltaic (PV) module safety qualification for DC system voltage up to 3000 V DC – Part 1: Requirements for construction • IEC TS 63546 ED1: Photovoltaic (PV) module safety qualification for DC system voltage up to 3000 V DC – Part 2: Requirements for testing.
IEC TC 34 [15]	Lighting	• **PWI 34-8:** LED light sources for horticultural lighting – Safety. • **PWI TR 34-12:** Environmental aspects for lighting products and systems • **PNW 34-1126 ED1:** Equipment for general lighting purposes – Objective light output flicker test method. • **PNW 34-1127 ED1:** Equipment for general lighting purposes – Objective Stroboscopic Effect Test Method. • **PNW 34-1151 ED1:** Excimer sources for germicidal purpose – Safety specifications. • **PNW 34-1152 ED1:** Measurement of ozone production from UV-C radiation sources and luminaires. • **PNW 34-1153 ED1:** Low-pressure mercury discharge UV radiation sources for germicidal purpose – Safety specifications. • **IEC 62386-105 ED2:** Digital addressable lighting interface – Part 105: Particular requirements for control gear and control devices – Firmware transfer. • **IEC 62386-302/AMD1 ED1:** Amendment 1 – Digital addressable lighting interface – Part 302: Particular requirements – Input devices – Absolute input devices. • **IEC 62386-303/AMD1 ED1:** Amendment 1 – Digital addressable lighting interface – Part 303: Particular requirements – Input devices – Occupancy sensor. • **IEC 62386-304/AMD1 ED1:** Amendment 1 – Digital addressable lighting interface – Part 304: Particular requirements – Input devices – Light sensor. • **IEC 62386-351 ED1:** Digital addressable lighting interface – Part 351: Particular requirements – Control devices – Luminaire-mounted control devices

(Continues)

		• **IEC 63128/AMD1 ED1:** Amendment 1 – Lighting control interface for dimming – Analogue voltage dimming interface for electronic current sourcing control. • **IEC 63403-1 ED1:** Horticultural lighting – LED packages for horticultural lighting – Part 1: Specification sheet • **IEC 63403-2 ED1:** Horticultural lighting – LED packages for horticultural lighting – Part 2: Binning • **IEC 63494-1 ED1:** Lighting System Electro-Mechanical Interfaces – Part 1: Safety • **IEC 63494-2-1 ED1:** Lighting system electro-mechanical interfaces – Part 2: Interchangeability requirements – Part 2-1: Four-pin ELV twist-lock interface – Type ZB18 • **IEC 63533 ED1:** Active airborne microorganism inactivation by Germicidal Ultraviolet (GUV) luminaires • **IEC 63535 ED1:** Germicidal UV luminaires – Radiation safety. • **IEC TR 63540 ED1:** Lighting systems – Characteristics for selected outdoor applications. • **IEC 63544 ED1:** Horti-luminaires incorporating LED sources – Performance. • **IEC 63545 ED1:** Horti-luminaires – Safety.
IEC TC 111 [16]	Environmental standardization for electrical and electronic products and system	• PNW 111-720 ED1: DETERMINATION OF CERTAIN SUBSTANCES IN ELECTROTECHNICAL PRODUCTS — Part 15: Tetrabromobisphenol A (TBBPA) in plastics by gas chromatography–mass spectrometry (GC–MS) and liquid chromatography–mass spectrometry (LC–MS) • IEC 62321-3-1 ED2: DETERMINATION OF CERTAIN SUBSTANCES IN ELECTROTECHNICAL PRODUCTS – Part 3-1: Elemental Screening by X-ray fluorescence spectrometry • IEC 62321-8 ED2: Determination of certain substances in electrotechnical products – Part 8: Phthalates in polymers by gas chromatography–mass spectrometry (GC–MS), gas chromatography–mass spectrometry using a pyrolyzer/thermal desorption accessory (Py-TD-GC–MS) • IEC 62321-10 ED2: Determination of certain substances in electrotechnical products – Part 10: Polycyclic aromatic hydrocarbons (PAHs) in polymers and electronics by gas chromatography–mass spectrometry (GC–MS)

(Continues)

Table 7.1 (Continued)

The TC	Responsibilities	The corresponding standards and codes
		• IEC 62321-13 ED1: technical products – Part 13: Bisphenol A in plastics by liquid chromatography–diode array detection (LC–DAD), liquid chromatography–mass spectrometry (LC–MS) and liquid chromatography–tandem mass spectrometry (LC–MS/MS) • IEC 62321-14 ED1: DETERMINATION OF CERTAIN SUBSTANCES IN ELECTROTECHNICAL PRODUCTS – Part 14: Determination of short-chain chlorinated paraffins (SCCPs) and medium-chain chlorinated paraffins (MCCPs) in electrotechnical products by gas chromatography–negative chemical ionization-mass spectrometry (GC–NCI-MS) • IEC 62635 ED1:Guidelines for end-of-life information provided by manufacturers and recyclers and for recyclability rate calculation of electrical and electronic equipment. • IEC 63366 ED1:Product category rules for life cycle assessment of electrical and electronic products and systems. • IEC 63372 ED1:Quantification and communication of Carbon FootPRINT and GHG emission reductions/avoided emissions from electric and electronic products and systems – Principles, methodologies, requirements, and guidance. • IEC 63395 ED1:Sustainable management of waste electrical and electronic equipment (e-waste). • IEC TS 63428 ED1: Guidance on material circularity considerations in environmentally conscious design • IEC 82474-1 ED1: Material declaration – Part 1: General requirements.
IEC TC 113 [17]	Nanotechnology for electrotechnical products and systems	• **IEC 62607-1**: Nano-enabled electrical energy storage devices. • **IEC 62607-2**: Nano-enabled energy conversion devices. • **IEC 62607-3**: Nano-enabled electronic devices. • **IEC 62607-4**: Nano-enabled sensors. • **IEC 62607-5**: Nano-enabled medical products. • **IEC 62607-6**: Nanomanufacturing – Key control characteristics – Part 6: Nano-enabled materials and other applications

(Continues)

IEC TC 44 [18]	Safety of machinery – Electrotechnical aspects	• **IEC 60204-1:** General requirements. • **IEC 60204-32:** Requirements for hoisting machines. • **IEC 60204-33:** Requirements for rubber and plastics mixing mills. • **IEC 60204-34**: Requirements for rolling mills and calendars in the rubber and plastics industries. • **IEC 60204-35:** Requirements for forming and printing machines used for the processing of plastics and rubber. • **IEC 60204-36:** Requirements for hoisting appliances.
IEC TC 47 [19]	Semiconductor devices	• **IEC 60749-34-1 ED1:** Semiconductor devices – Mechanical and climatic test methods – Part 34-1: Power cycling test for power semiconductor module • **IEC 63150-2 ED1:** Semiconductor devices – Measurement and evaluation methods of kinetic energy harvesting devices under practical vibration environment – Part 2: Human arm swing motion. • **IEC 63150-3 ED1:** Semiconductor devices – Measurement and evaluation methods of kinetic energy harvesting devices under practical vibration environment – Part 3: Human foot impact motion. • **IEC 63287-3 ED1:** Semiconductor devices – Generic semiconductor qualification guidelines – Part 3: Guidelines for reliability qualification plans for power semiconductor module. • **IEC 63419 ED1**: Guideline for Switching Reliability Evaluation Procedures for Gallium Nitride Power Conversion Devices • **IEC 63492-1 ED1:** Semiconductor devices – Isolation for semiconductor devices – Part 1: Failure mechanisms and measurement methods to evaluate solid insulation for semiconductor devices. • **IEC 63505 ED1:** Guidelines for measuring the threshold voltage (VT) of SiC MOSFETs • **IEC 63512 ED1:** Test method for continuous-switching evaluation of gallium nitride power conversion devices.

(Continues)

Table 7.1 (*Continued*)

The TC	Responsibilities	The corresponding standards and codes
IEC TC 22 [20]	Power electronic systems and equipment	• **IEC 60146-1-1 ED5:** Semiconductor converters – General requirements and line commutated converters – Part 1-1: Specification of basic requirements. • **IEC 62477-2 ED2**: Safety requirements for power electronic converter systems and equipment – Part 2: Power electronic converters from 1000 V AC or 1500 V DC up to 36 kV AC or 54 kV DC. • **IEC TS 63490 ED1**: Terms and Definition for standards incorporating power electronic conversion.
TC 110 [21]	Electronic displays	• **IEC 62341-6-1 ED4:** Organic light emitting diode (OLED) displays – Part 6-1: Measuring methods of optical and electro-optical parameters. • **IEC 62341-6-7 ED1:** Organic light emitting diode (OLED) displays – Part 6-7: Measuring methods of optical characteristics for display with the under-screen feature. • **IEC 62595-1-2 ED3**: Display lighting unit – Part 1-2: Terminology and letter symbols • **IEC TR 62595-1-6 ED1:** Display light unit – Part 1-6: Quantum dot films and quantum dot diffuser plates used in the backlight unit. • **IEC TR 62629-1-3 ED1:** IEC/TR 62629-1-3 ED1: 3D display devices –Part 1-3: Depth perception by humans and determination of the position of a 3D object on the non-physical screen • **IEC 62629-52-1 ED1:** 3D displays – Part 52-1: Fundamental measurement methods of aerial display – Optical • IEC 62629-62-12 ED1: 3D displays – Part 62-12: Measurement methods for virtual-image type – Image Quality • IEC TR 62715-5-61 ED1: Flexible display devices – Part 5-61: Stretchable displays – Overview of measurement and application scenarios • IEC 62715-6-23 ED1: Flexible display devices – Part 6-23: Mechanical misaligned folding test method • IEC TR 62715-6-41 ED1: Flexible display devices – Part 6-41: General introduction for rollable displays • IEC 62715-6-42 ED1: Flexible display devices – Part 6-42: Flattening force measurement methods

(Continues)

- IEC 62906-5-8 ED1: LASER DISPLAYS – Part 5-8: Measurement of scanning characteristics for raster-scanning laser display
- IEC 62906-6-1 ED1: LASER DISPLAYS – Part 6-1: Visualization method of color gamut intersection
- IEC 62908-22-10 ED1: Touch and interactive displays – Part 22-10: Measuring methods of fingerprint recognition performance – Under-display optical imaging fingerprint sensing
- IEC 62908-42-10 ED1: Touch and interactive displays – Part 42-10: Measurement methods of motion-tracking image-control response time for interactive projection display
- IEC 62977-2-7 ED1: Electronic displays – Part 2-7: Measurements of optical characteristics – Tiled displays
- IEC 62977-2-8 ED1: Electronic Displays – Part 2-8: Measurements of optical characteristics – Reflective displays
- IEC 62977-3-6 ED1: Electronic displays – Part 3-6: Evaluation of optical performance – Spatial resolution
- IEC 63145-20-10 ED2: Eyewear display – Part 20-10: Fundamental measurement methods – Optical properties
- IEC 63145-22-20 ED1: Eyewear display – Part 22-20: Specific measurement methods for AR type – Image quality
- IEC TR 63145-40-20 ED1: EYEWEAR DISPLAY – Part 40-20: 3D sensing
- IEC 63145-201-10 ED1: Eyewear display – Part 201-10: Measurement methods for VR type – Optical properties of a singlet lens used for eyepieces.
- IEC TR 63145-202-40 ED1: Eyewear display – Part 202-40: General information of AR type – Frontal stray lights
- IEC 63211-2-21 ED1: Durability test methods for electronic displays – Part 2-21: Environmental tests – Test methods for heat and humidity
- IEC 63211-2-23 ED1: Durability test methods for electronic displays – Part 2-23: Environmental tests – Outdoor weathering.

(Continues)

Table 7.1 (*Continued*)

The TC	Responsibilities	The corresponding standards and codes
		• IEC 63211-3-2 ED1: Durability test methods for electronic displays – Part 3-2: Mechanical tests – Static stress. • IEC TR 63340-1 ED1: Electronic displays for special applications – Part 1: General introduction. • IEC TR 63340-2 ED1: Electronic displays for special applications – Part 2: Elevator and escalator. • IEC TR 63340-3 ED1: Electronic displays for special applications – Part 3: Gaming and e-sports
TC 61 [22]	Safety of household and similar electrical appliances	• **IEC 60335 series:** Safety of household and similar electrical appliances • **IEC 60364 series:** Electrical installations of buildings. • **IEC 60950 series:** Information technology equipment. • **IEC 60445:** Basic and safety principles for man-machine interface, marking, and identification. • **IEC 62040-3:** Uninterruptible power systems (UPS).
TC 70 [23].	Degrees of protection provided by enclosures	• **IEC 60529:** Degrees of protection provided by enclosures (IP Code). • **IEC 62262:** Degrees of protection provided by enclosures for electrical equipment against external mechanical impacts. • **IEC 61032:** Protection of persons and equipment by enclosures.
IEC TC 79 [24]	Alarm and electronic security systems	• **IEC 62676-2-11 ED1:** Alarm systems – Video Surveillance Systems (VSS) for use in security applications – Part 2-11: Video transmission protocols – Interop profiles for VMS- and cloud VSaaS-systems for safe-cities and law-enforcement • **IEC 62676-4 ED2:** Video surveillance systems for use in security applications – Part 4: Application guidelines. • **IEC 62676-5 ED2:** Video surveillance systems for use in security applications – Part 5: Data specifications and image quality performance for camera devices.

(Continues)

		• **IEC 62676-5-1 ED1:** Video surveillance systems for use in security applications – Part 5-1: Data specifications and image quality performance for camera devices – Environmental test methods for image quality performance. • **IEC 62676-6 ED1:** Video surveillance systems for use in security applications – Part 6: Performance testing and grading of real-time intelligent video content analysis devices & systems for use in video surveillance applications. • **IEC 62820-1-1 ED2:** Building intercom systems – Part 1-1: System requirements – General. • **IEC 62820-1-2 ED2:** Building intercom systems – Part 1-2: System requirements – Building intercom systems using the internet protocol (IP).
IEC TC 64 [25]	Electrical installations and protection against electric shock	• IEC 60364-1 ED6: Low-voltage electrical installations – Part 1: Fundamental principles, assessment of general characteristics, definitions • IEC 60364-4-41 ED6:Amendment 2 – Low-voltage electrical installations – Part 4-41: Protection for safety – Protection against electric shock • IEC 60364-4-42 ED4:Low-voltage electrical installations – Part 4-42: Protection for safety – Protection against thermal effects • IEC 60364-4-44/AMD3 ED2:Amendment 3 – Low-voltage electrical installations – Part 4-44: Protection for safety – Protection against voltage disturbances and electromagnetic disturbances – Clause 443 • IEC 60364-5-52/AMD1 ED3:Amendment 1 – Low-voltage electrical installations – Part 5-52: Selection and erection of electrical equipment – Wiring systems • IEC 60364-5-53/AMD2 ED4:Amendment 2 – Low-voltage electrical installations – Part 5-53: Selection and erection of electrical equipment – Devices for protection for safety, isolation, switching, control and monitoring • IEC 60364-6 ED3: Low voltage electrical installations – Part 6: Verification • IEC 60364-7-702 ED4: Low-voltage electrical installations – Part 7-702: Requirements for special installations or locations – Swimming pools and fountains

(Continues)

Table 7.1 (*Continued*)

The TC	Responsibilities	The corresponding standards and codes
		• IEC 60364-7-706 ED3: Low-voltage electrical installations – Part 7-706: Requirements for special installations or locations – Conducting locations with restricted movement. • IEC 60364-7-711 ED3: Low-voltage electrical installations – Part 7-711: Requirements for special installations or locations – Temporary electrical installations for exhibitions and entertainment-related purposes • IEC 60364-7-712 ED3: Low voltage electrical installations – Part 7-712: Requirements for special installations or locations – Solar photovoltaic (PV) power supply systems • IEC 60364-7-717 ED3: Low-voltage electrical installations – Part 7-717: Requirements for special installations or locations – Mobile or transportable units • IEC 60364-7-725 ED1: IEC 60364 Low-voltage electrical installations – Part 7-725: Requirements for special installations or locations – Resilient power supply system • IEC 60364-7-751 ED1: Low-voltage electrical installations – Part 7-751: Requirements for special installations or locations – Low voltage generating. • IEC 60479-2 ED2: Effects of current on human beings and livestock – Part 2: Special aspects. • IEC TR 60479-5 ED2Effects of current on human beings and livestock – Part 5: Touch voltage threshold values for physiological effects. • IEC TR 61200-201 Application guides complying with IEC 60364 – Asynchronous motor starting and protection. • IEC TR 61200-202 ED1: Application guides complying with IEC 60364 – Lighting circuits. • IEC TR 61200-203 ED1: Application guides complying with IEC 60364 – Uninterruptible Power Systems • IEC TR 61200-204 ED1: Application guides complying with IEC 60364 – Rotating generators. • IEC TR 61200-205 ED1: Application guides complying with IEC 60364 – Source changeover system.
IEC TC 89 [26]	Fire hazard testing	• **IEC TS 60695-2-20 ED4:** Fire hazard testing – Part 2-20: Glowing/hot-wire based test methods – Hot wire coil test method – Apparatus, verification, test method and guidance.

(Continues)

IEC TC 8 [27]	System aspects of electrical energy supply	• IEC TR 63282-101 LVDC systems: DC power distribution system for typical scenarios • IEC TR 63282-102 LVDC systems: Technical Report for low-voltage DC independent power supply system • IEC TR 62786-100 Distributed energy resources connection with the grid – Part 100 Generating units grid connection standard mapping • IEC TR 62786-101 Distributed energy resources connection with the grid – Part 101 Gravity storage connection to the grid • IEC TR 62786-102 Distributed energy resources connection with the grid – Part 102 CAES connection to the grid • IEC TR 62786-103 Distributed energy resources connection with the grid – Part 103 V2G application scenarios for the grid • PNW TS 8-1683 ED1: Distributed energy resources connection with the grid – Part 2 Additional requirements for PV generation • **IEC 60038/AMD2/FRAG1 ED7:** Amendment 2 (Fragment 1) – Standard voltages for LVDC supply and LVDC equipment (Proposed horizontal standard) • **IEC 60038/AMD2/FRAG2 ED7:** Amendment 2 (Fragment 2) – Standard voltages for HVDC supply and HVDC equipment (Proposed horizontal standard) • **IEC 60050-602/AMD1 ED1:** Amendment 1 – International Electrotechnical Vocabulary (IEV) – Part 602: Generation, transmission and distribution of electricity – Generation • **IEC 60050-603/AMD2 ED1:** Amendment 2 – International Electrotechnical Vocabulary (IEV) – Part 603: Generation, transmission, and distribution of electricity – Power systems planning and management. • **IEC 60050-605/AMD1 ED1:** Amendment 1 – International Electrotechnical Vocabulary (IEV) – Part 605: Generation, transmission and distribution of electricity – Substations • **IEC 60050-614/AMD1 ED1:** Amendment 1 – International Electrotechnical Vocabulary (IEV) – Part 614: Generation, transmission and distribution of electricity – Operation

(Continues)

Table 7.1 (*Continued*)

The TC	Responsibilities	The corresponding standards and codes
		• IEC 60050-617 ED2: International Electrotechnical Vocabulary (IEV) – Part 617: Organization/Market of electricity. • **IEC 60050-691/AMD2 ED1:** Amendment 2 – International Electrotechnical Vocabulary (IEV) – Part 691: Tariffs for electricity • **IEC TS 62749 ED3**: Assessment of power quality – Characteristics of electricity supplied by public networks. • **IEC TS 62786-42 ED1:** Distributed energy resources connection with the grid – Part 42 Requirements for voltage measurement used to control DER and loads. • **IEC TS 63222-1 ED2:** Power quality management – Part 1: General guidelines • **IEC TS 63222-3 ED1:** Power quality management – Part 3: User characteristics modelling. • IEC TS 63222-4 ED1: Power quality management – Part 4: Harmonic analysis over public supply network • IEC TR 63222-101 ED1: IEC TR 63222-101 Power quality management – Part 101: Power quality data application • IEC TR 63282 ED2: LVDC systems – Assessment of standard voltages and power quality requirements.
IEC TC 77 [28]	Electromagnetic compatibility	• **PNW** TS 77-596 ED1: Electromagnetic compatibility (EMC) – Part 1-6: General – Guide to the evaluation of measurement uncertainty • IEC TR 61000-4-1 ED2: Electromagnetic compatibility (EMC) – Part 4-1: Testing and measurement techniques – Overview of IEC 61000-4 series

- IEC TC 2 is responsible for the rotating machinery and sets of IEC 60034 series [10]
- IEC TC 105 sets the standards of fuel cell technologies (IEC 62282-4-101) [11].
- IEC TC 40 is responsible for setting the standards for electric double-layer capacitors (IEC 62391-1) [12].
- IEC TC 23 is responsible for setting the standards for the electrical accessories (IEC 62613) including socket outlets, plugs, and ship couplers for high-voltage shore connection units as well as it sets low-voltage shore connection systems (IEC 60309-5) [13].
- IEC TC 82 is responsible for setting the standards for the solar photovoltaic panel (IEC 62788) [14].
- IEC TC 34 is responsible for setting the standards for lamps as well as the related equipment (IEC 62031 the safety requirements for LEDs) [15].
- IEC TC 111 is responsible for setting the standards for the environmental standardization of electronic and electrical products and systems; this committee sets two standards including the levels of potentially toxic substances in electrical and electronic products (EC 62321) [16].
- The standards for reducing the adverse environmental impact of devices throughout their lifecycle (IEC 62430).
- IEC TC 113 is responsible for setting the standards of nanotechnology for electrotechnical products and systems [17].
- IEC TC 44 is responsible for setting the standards of the electrotechnical aspects of machinery safety [18].
- IEC TC 47 is responsible for setting the standards for semiconductor devices [19].
- IEC TC 22 is responsible for setting the standards of power electronic equipment and systems [20].
- IEC TC 110 is responsible for setting the standards of electronic display devices, this committee already set several standards for virtual reality (VR), AR, and MR displays (IEC 63145-1-2) [21].
- IEC TC 61 is responsible for setting the standards of safety of households as well as similar electrical appliances (IEC 60335-2-90) [22].
- IEC TC 70 is responsible for setting the standards of the degrees of protection provided enclosures (IEC 60529) [23].
- IEC TC 79 is responsible for setting the standards of Alarm and electronic security systems and it published the IEC 60747-14-1 for sensor specification and IEC 62676-1-1 for the CCTV requirements [24].
- ISO/IEC SC 4179 is a subcommittee that is responsible for setting the standards of the Internet of Things and digital twin [29].
- IEC TC 64 is responsible for setting the standards of electrical installations and protection against electric shock (IEC 61140) [25].
- IEC TC 89 is responsible for setting the standards of the fire hazard testing (IEC 60695-1-10) [26].
- IEC TC 8 is responsible for setting the standards of system aspects of electrical energy supply [27].

- IEC TC 77 is responsible for setting the standards of system aspects of electromagnetic compatibility and it sets IEC 61000 which terminology, descriptions of EM phenomena and the electromagnetic compatibility (EMC) environment [28].

IMO has established guidelines on maritime cyber risk management, referring to ISO/IEC 27001 which specifies requirements for implementing and maintaining a security management system. The standard is part of the ISO/IEC 27000 family, a critical set of publications dealing with cyber security, published by ISO/IEC JTC 1/SC 27.

7.4 The IEC conformity assessment (CA)

CA are systems or mechanisms that are used to ensure that the electric systems in maritime transport are followed and fulfilled [30]. To ensure that the systems and equipment meet the standard equipment, the CA involves a set of activities including testing and measurement, inspection, certification, supplier's declaration of conformity (SDOC), and accreditation (Figure 7.2).

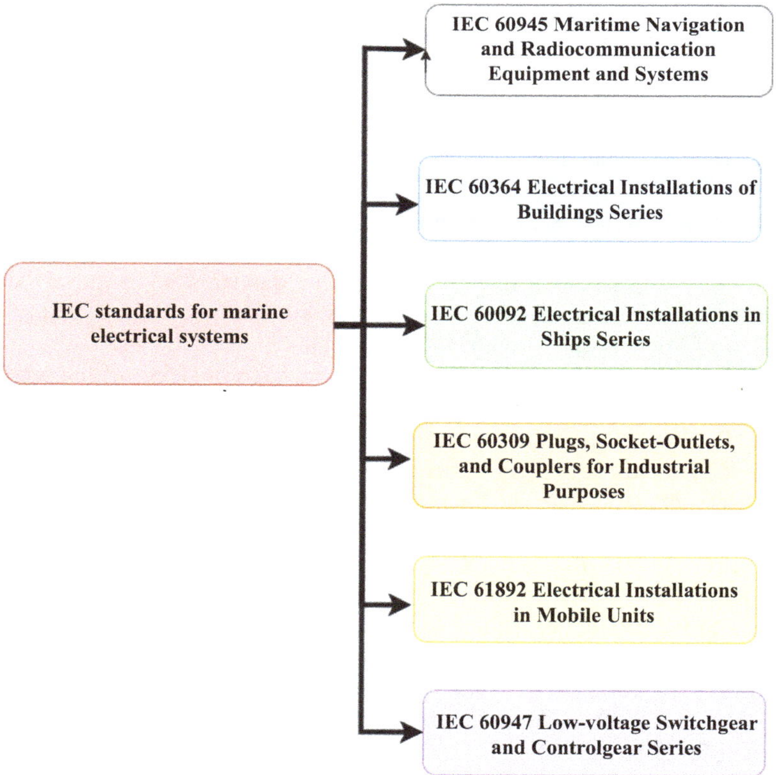

Figure 7.2 The IEC standard for marine electric systems

References

[1] "Electric ship market," https://www.marketsandmarkets.com/Market-Reports/electric-ships-market-167955093.html (accessed Jan 2025).

[2] "Global electric ship market," https://www.statista.com/statistics/1097195/projected-size-of-the-global-electric-ship-market/ (accessed Jan 2025).

[3] "The National Electrical Code (NEC)," https://www.esfi.org/workplace-safety/industry-codes-regulations/the-national-electrical-code-nec/ (accessed Jan 2025).

[4] "NFPA 70," https://www.nfpa.org/codes-and-standards/7/0/nfpa-70 (accessed Jan 2025).

[5] "The American Boat and Yacht Council (ABYC)," https://abycinc.org/ (accessed Jan 2025).

[6] "The International Maritime Organization," https://www.imo.org/ (accessed Jan 2025).

[7] "TC 80: Maritime navigation and radiocommunication equipment and systems," https://www.iec.ch/dyn/www/f?p=103:7:0::::FSP_ORG_ID,FSP_LANG_ID:1271,25 (accessed Jan 2025).

[8] "TC 18: Electrical installations of ships and of mobile and fixed offshore units," https://www.iec.ch/dyn/www/f?p=103:7:0::::FSP_ORG_ID,FSP_LANG_ID:1284,25 (accessed Jan 2025).

[9] "Technical committees and subcommittees," https://www.iec.ch/technical-committees-and-subcommittees#tclist (accessed Jan 2025).

[10] "TC 2: Rotating machinery," https://iec.ch/dyn/www/f?p=103:7:0::::FSP_ORG_ID,FSP_LANG_ID:1221,25 (accessed Jan 2025).

[11] "TC 105: Fuel cell technologies," https://www.iec.ch/dyn/www/f?p=103:7:0::::FSP_ORG_ID,FSP_LANG_ID:1309,25 (accessed Jan 2025).

[12] "TC 40: Capacitors and resistors for electronic equipment," https://www.iec.ch/dyn/www/f?p=103:7:0::::FSP_ORG_ID,FSP_LANG_ID:1219,25 (accessed Jan 2025).

[13] "TC 23: Electrical accessories," https://iec.ch/dyn/www/f?p=103:7:0::::FSP_ORG_ID,FSP_LANG_ID:1299,25 (accessed Jan 2025).

[14] "TC 82: Solar photovoltaic energy systems," https://www.iec.ch/dyn/www/f?p=103:7:0::::FSP_ORG_ID,FSP_LANG_ID:1276,25 (accessed Jan 2025).

[15] "TC 34: Lighting," https://www.iec.ch/dyn/www/f?p=103:7:0::::FSP_ORG_ID,FSP_LANG_ID:1235,25 (accessed Jan 2025).

[16] "TC 111: Environmental standardization for electrical and electronic products and systems," https://www.iec.ch/dyn/www/f?p=103:7:0::::FSP_ORG_ID,FSP_LANG_ID:1314,25 (accessed Jan 2025).

[17] "TC 113: Nanotechnology for electrotechnical products and systems," https://www.iec.ch/dyn/www/f?p=103:7:0::::FSP_ORG_ID,FSP_LANG_ID:1315,25 (accessed Jan 2025).

[18] "TC 44: Safety of machinery – Electrotechnical aspects," https://www.iec.ch/dyn/www/f?p=103:7:0::::FSP_ORG_ID,FSP_LANG_ID:1302,25 (accessed Jan 2025).
[19] "TC 47: Semiconductor devices," https://iec.ch/dyn/www/f?p=103:7:0::::FSP_ORG_ID,FSP_LANG_ID:1251,25 (accessed Jan 2025).
[20] "TC 22: Power electronic systems and equipment," https://www.iec.ch/dyn/www/f?p=103:7:0::::FSP_ORG_ID,FSP_LANG_ID:1293,25 (accessed Jan 2025).
[21] "TC 110: Electronic displays," https://www.iec.ch/dyn/www/f?p=103:7:0::::FSP_ORG_ID,FSP_LANG_ID:1313,25 (accessed Jan 2025).
[22] "TC 61: Safety of household and similar electrical appliances," https://www.iec.ch/dyn/www/f?p=103:7:0::::FSP_ORG_ID,FSP_LANG_ID:1236,25 (accessed Jan 2025).
[23] "TC 70: Degrees of protection provided by enclosures," https://www.iec.ch/dyn/www/f?p=103:7:0::::FSP_ORG_ID,FSP_LANG_ID:1256,25 (accessed Jan 2025).
[24] "TC 79: Alarm and electronic security systems," https://www.iec.ch/dyn/www/f?p=103:7:0::::FSP_ORG_ID,FSP_LANG_ID:1269,25 (accessed Jan 2025).
[25] "TC 64: Electrical installations and protection against electric shock," https://iec.ch/dyn/www/f?p=103:7:0::::FSP_ORG_ID:1249 (accessed Jan 2025).
[26] "TC 89: Fire hazard testing," https://www.iec.ch/dyn/www/f?p=103:7:0::::FSP_ORG_ID,FSP_LANG_ID:1283,25 (accessed Jan 2025).
[27] "TC 8: System aspects of electrical energy supply," https://www.iec.ch/dyn/www/f?p=103:7:::::FSP_ORG_ID:1240 (accessed Jan 2025).
[28] "TC 77: Electromagnetic compatibility," https://www.iec.ch/dyn/www/f?p=103%3A7%3A0%3A%3A%3A%3AFSP_ORG_ID%2CFSP_LANG_ID%3A1265%2C25 (accessed Jan 2025).
[29] "ISO/IEC JTC 1/SC 41: Internet of things and digital twin," https://www.iso.org/committee/6483279.html (accessed Jan 2025).
[30] "The IEC Conformity Assessment (CA)," https://www.iec.ch/conformity-assessment/what-conformity-assessment (accessed Jan 2025).

Chapter 8

Digitalization (digital twin) impact on transportation applications

Adel El-Shahat[1], Zain K. Mirza[1], Jason Yu-Chen Qian[1] and Ziad M. Ali[2]

This chapter explores how the use of digitalization can solve transportation issues (how to efficiently manage transport flows) using intelligent transport system (ITS) implementation. ITS are smart systems that use innovative developments in modeling transport systems. The aim is to provide users with more information, content, and safety. This would include city services and traffic police.

The aim of implementing transport control centers would aim to achieve the following tasks:

- Efficient traffic flow
- Monitoring road networks
- Coordination of resources in emergency situations
- Informing individuals of alternative routes in emergency situations
- Decreasing consequences in emergency situations

Some of the problems relating to transport systems:

Objective problems	Subjective problems
• Rising vehicle to population ratio • Increased use of individual transportation • Decreased passenger transport efficiency • Increased need for urban area residents to move • Dispersion in ratio of vehicle to population and progression of road construction • Issues in urban planning and urban area development	• Lacking systems for managing complex road transportation • Insufficient legislative base for management of city transport system • Insufficient funding of the development of transportation infrastructure • Property rights issue • Negative impact on human factors

[1]School of Engineering Technology, Purdue University, USA
[2]College of Engineering at Wadi Addawasir, Prince Sattam Bin Abdulaziz University, KSA

How digital twin (DT) is used in this process:

- DT uses mathematical modeling methods to analyze transportation networks and then suggest solutions based on the analysis.
- A transportation DT would be used to generate a digital version of a transportation object or process, providing a real-time data exchange between the physical object/system and the DT.
    - This source introduces the use of DT for adaptive traffic signal control (ATSC) to improve transportation by reducing waiting time at intersections.

Figure 8.1 shows a proposed DT implementation in traffic signals. Vehicle trajectories and signal data are transmitted from the physical twin using both wired and wireless communication. Using this information, the DT will then send appropriate signal phases and timing parameters in real time to notify the physical systems about the findings.

Some challenges for fully adopting a DT system:

1. Interoperability
2. Authentication
    (a) Efficient technology needed that will enable real-time data exchange and analysis on multiple levels and various environments.

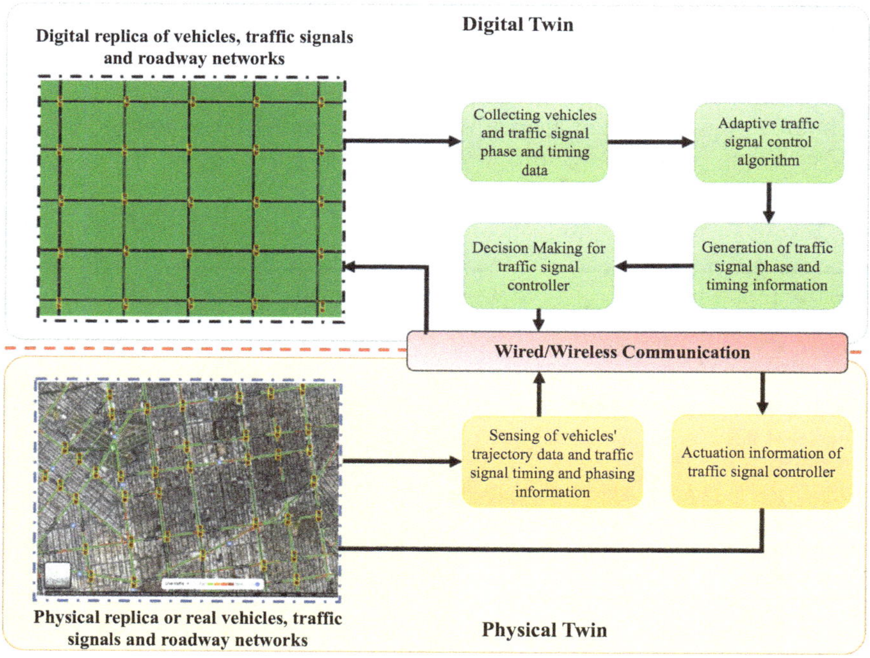

Figure 8.1 Digital twin in traffic signal control signals

(b) Large amounts of data must be collected and analyzed prior to building the DT in order to make accurate predictions.
(c) Systems must be able to accommodate multiple DTs that include multiple factors involved including objects, devices, and people. The system must be able to deal with DTs operating simultaneously.

3. Distributed machine learning
4. Distributed decision-making
5. Scalability and robustness

The DT will help understand what is simultaneously happening on the physical level while predicting what may happen in the future. In order to get a properly functioning DT, there are a set of requirements that must be fulfilled, listed as follows:

List of requirements:

1. Data collection
   - Reasoning: to support data-driven decision-making
2. Data update frequency
   - Reasoning: to provide simultaneous updates of physical twin
3. Data analysis
   - Reasoning: Enable advanced predictions of any possible risks
4. Simulation capabilities
   - Reasoning: Enable virtual visibility
5. Data exchange
   - Reasoning: Enable data sharing from physical twin to DT
6. Authentication
   - Reasoning: Maintain trust and security among network
7. Transparency
   - Reasoning: Allow traceability across network
8. Distribute decision-making capabilities
   - Reasoning: Provide a consensus-based decision-making process
9. Decentralization
   - Reasoning: Provide efficient and reliable solutions

This case study investigates the impact of digitalization on air transportation systems (ATS) in several countries, including Australia, China, the USA, and India. Air transportation is defined by three elements which include air traffic control, airport, and airlines. This research first analyzes the different types and levels of digitalization that can occur in air transportation, then it looks at the impact that digitalization might lead to, and lastly they use forecasts to predict how digitalization will look in the next coming years in the countries mentioned.

8.1 Introduction

Aside from the three components of ATS mentioned above, one must consider the human, political, regulation, economic, environmental, and technological factors that are important aspects to consider when conducting research related to ATS. As demand increases and networks become more complex, there is a need for transport systems to keep up pace. Since digitalization and intelligentization play a vital role in improving the efficiency of ATS, the research on the influence of digitalization and intelligentization on ATS is significant.

8.2 Methodology

The data obtained from this resource come from three main sources: World bank, international organizations websites, and national aviation reports. There were four main methodologies used to analyze data and examine how digitalization might affect ATS in the sample countries listed.

1. Comparative analysis

   Used to compare each case with related data in the nature of the cases. Comparative analysis includes selection of comparison cases, measurement, linkage and gap in knowledge, failed data and cases, time dimension, and selection of other relative methods. Using these three steps, the comparative analysis method can be applied:
   (a) Establish objective of investigation
   (b) Listing and plot data
   (c) Analyze and make conclusions based on data

   In terms of this chapter, the comparative analysis model is used to look at the different levels of digitalization that can be applied to the ATS.
2. Correlation analysis
   - Statistical method that studies the correlation of more than two sets of variables.
   - In terms of this chapter, correlation analysis is used to study the correlation of the different impact factors with the ATS in different countries.
3. Regression analysis
   - Statistical method that explores the functional relationship among variables.
   - An equation model is used to express the effect of couples of independent variables on a dependent variable.
   - Regression analysis will be used to analyze the relationship between national ATS with the development of digitalization.
4. Forecasting analysis
   - Used to predict the future trend of the objectives based on the historical data and the present data.

- Uses regression analysis to make a functional relationship model
- In this research, it is used to predict the future trend of digitalization and intelligentization's impact on ATS in the sample countries in the next few years.

8.3 Results

Using comparative analysis, the number of air transport passengers carried is shown to be increasing year by year in the four sample countries used in this research. The USA has the highest numbers which can be related to a highly developed aviation industry. Figure 8.2 shows the comparison between all four countries in the study.

Figures 8.3 and 8.4 show further results of the analysis that was done on the sample countries on the amount of high-tech export and digital comparative rankings.

Using the analysis results, it was evident that digitalization would have different impact levels on the four sample countries. This means that digitalization and intelligentization have an impact on the development of ATS in Australia. For China, digitalization and intelligentization have the most impact on the development of ATS. The results of the analysis showed that digitalization has very little influence on the frequency and levels of air traffic accidents.

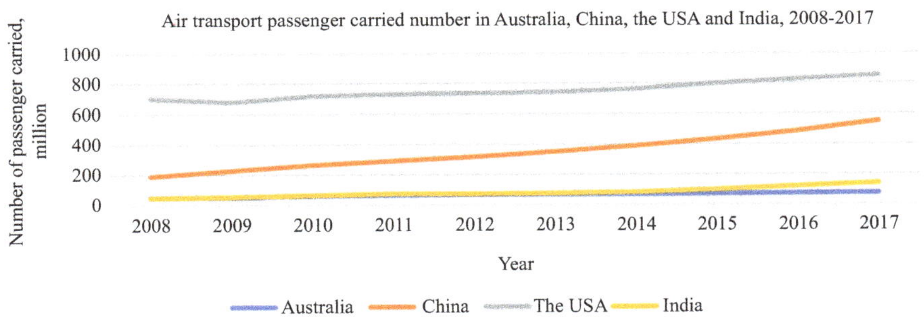

Figure 8.2 Air transport passenger carried number in sample countries, numbers in million

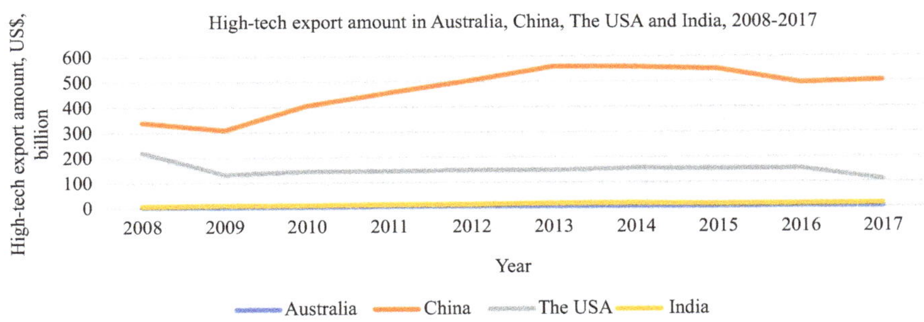

Figure 8.3 High-tech export amount in sample countries, US$ in billion

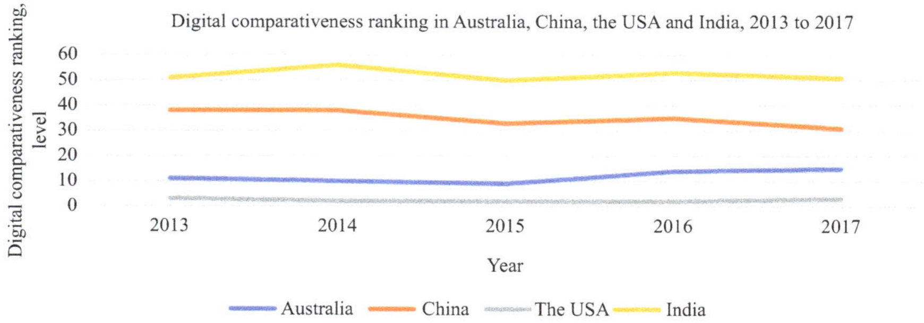

Figure 8.4 Digital comparativeness ranking in sample countries

8.4 Implications

Using the analysis in this study, digitalization in ATS could be found to help develop investment plans and improvement strategies. The four countries in the study included both developed and developing countries emphasizing the importance of development levels of digitalization in ATS. The correlation analysis identifies the most impact factors which need to be focused on to improve the different aspects of ATS.

A DT is a virtual copy or representation of an asset or a process which serves the purpose of a real-time digital replica of the same.

Figure 8.5 is an example using an overview of an aircraft; the key is once the DT is realized, it must be fed real-time information allowing for a safe environment to analyze its behavior and those of its components.

By doing these developments, it allows for an exact and safe comprehensive analysis of any asset and its components under normal usage and wear.

DT are of two types:

1. DT prototype
2. DT instance

All DT developments are executed in a digital twin environment.

Some of the key aspects of DT development are the connection and collection of information linked to the asset.

Some of the benefits of DT development are as follows:

DT has been one of the most significant trends of what we have come to call the fourth industrial revolution thanks to the impact on industrial manufacturing due to the remarkable progress in communication and information technologies.

DT has impacted various fields, but this paper specifically talks about the impact on electric vehicle (EV) propulsion drive systems.

DT trend began with computer simulation, but it rapidly showed some drawbacks such as mistakes in the programming, limited time to analyze results, and lack of data exchange between the real and the simulation models, among others [2].

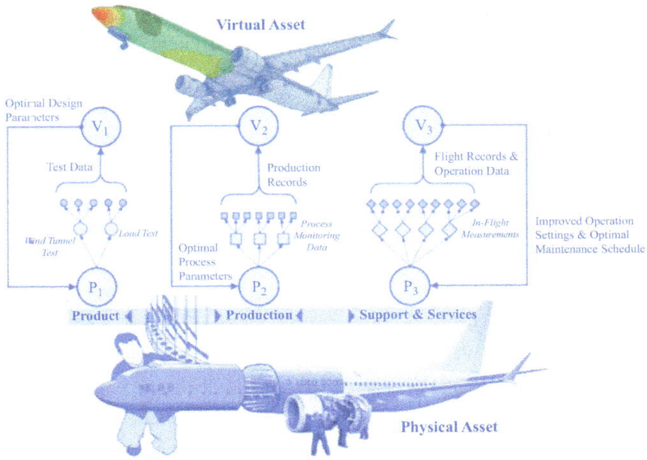

Figure 8.5 Representation of the DT concept

As a result, more real-time simulation tools are required to produce the hardware-in-the-loop (HIL) system. Under this technique, physical signals from a controller are connected to a test system that physically simulates the situation, tricking the controller into thinking it is in the assembled product. This technique however also presented some drawbacks due to the complexity of development and verification.

DTs, also known as mega computational models, can be defined as a digital representation of a physical object, process, or service. In essence, it is a computer program that uses real-world data to create simulations that can predict how a product or process will perform. These programs integrate IoT (Internet of Things), artificial intelligence, and software analytics.

The chapter explains that one of the main advantages between DT and simulations is that DT can study more issues for more vantage points due to the added computing power that accompanies a virtual environment. As represented in Table 8.1, the comparisons between computer simulations, HIL, and DT show that DTs receive real-time updates from the physical asset, process, or system.

Table 8.1 Comparison between computer simulation, HIL, and DT

	Computer simulation	**HIL**	**DT**
Data element and interaction	Static	Active	Active
Simulation basis	Potential parameters	Real-time feedback	Real-time feedback
Scope	Narrow-primary design	Narrow-advanced	Wide-advanced

The chapter also explains the application of DT technology in various sectors such as aerospace, automotive, healthcare, smart cities, construction, and automation. We have seen the development of EV propulsion drive systems within this industry. These systems have been challenging to design. DTs for EV drive systems are frequently used for system health monitoring and diagnostics, but it is still a new topic that is fully covered by DT technology. According to the authors, DTs for EV propulsion drive systems are recommended to be applied in designing or testing the system's performance. However, DT may be more effective once the physical drive system model is obtained.

As a result of machine learning and Big Data advancements, these virtual models have become very significant in modern engineering to drive innovation and improve performance. We will continue to see the development of computer simulations, and specifically DT in various fields. Future work will include further comparisons and applications of HIL simulations and DTs for the application of EV propulsion drive system design, depending on the stage of the work (Table 8.2).

Digital twin definition: "A digital twin is an instance of a digital model that represents an entity, which includes assets, processes, and systems which are sufficient to meet the requirements of a set of use cases."

8.5 Explanation of definition

Digital model: the model describes the different characteristics and behaviors of an entity.

Instance: creates a unique DT for each entity, describes how the model contains the data and where the information lives. The model defines what a DT's structure is, but still the twin is one stand-alone instance for an entity.

Industrial DT categories:

Discrete DT

- The lowest level but will be enough to meet certain requirements of a use case. A single entity adds no value as it does not need to be broken down further since the status of it is seen at entity level.

Composite DT

- A combination of several discrete DTs that represent an entity made up of several smaller parts.

The benefits of DTs:

1. Sharing data across silos

   Data can be shared across different silos in an organization. In the past, organizations have managed their data in their own storage mechanisms. The DT would be able to collect data from several sensors while simultaneously understanding the relationship between the sensors.

Table 8.2 Comparison of existing technologies for smart transportation

Highlighted	Blockchain	DT	Collaboration	Data analysis	IoT	Selected use cases logistics	Railway
MVDF technique for handling asynchronous generated from the connected vehicles in STS	X	X	X	✓	✓	X	X
Enhancing the security and performance of nodes on the Internet of Vehicles.	X	X	X	X	✓	X	X
Hybridized cryptographic integrated steganography algorithm for secured data sharing in IoT in cloud environment	X	X	X	X	✓	X	X
IoT-assisted intelligent logistics transportation management framework to optimize logistics	X	X	X	X	✓	✓	X
Exploring potential of using DT in synchromodal transport	X	/	X	X	✓	X	X
Using DTs to apply maintenance strategies in the industrial sectors.	X	/	X	X	✓	X	X
Identify a set of requirements to enable predictive maintenance for Industry 4.0 including railway.	X	X	X	X	✓	X	✓
Exploring and analysis of adoption of blockchain in the railway	✓	X	X	X	X	X	✓

(Continues)

Table 8.2 (Continued)

Highlighted	Blockchain	DT	Collaboration	Data analysis	IoT	Selected use cases logistics	Railway
Proposing a DT-based maintenance model for cranes operating in control terminal.	X	✓	X	X	✓	✓	X
Proposing multi-component based on DTS called Ilmatar for overhead cranes.	X	✓	X	X	✓	✓	X
Proposing a robotic process automation solution to deliver service for the patients	X	✓	X	✓	X	✓	X
Study the potential of using DTs to manage the COVID-19	X	✓	X	✓	X	X	X
A blockchain framework for heterogeneous multi-robot collaboration to combat COVID-19	✓	X	✓	X	✓	X	X
The collaboration between drone and IoT devices for improving Industry 4.0 applications such as smart city, smart healthcare	✓	X	✓	X	✓	X	X

2. Managing operations in real time

   DTs have the capability to create situational awareness by compiling a "bigger picture" of the entity it represents. Not only does it store data, but it also uses real-time operations by using visibility and creating decisions based on the data.
3. Conduct simulations and experiments

   DTs have the capability to make simulations of situations that allow the user to experiment on the twin rather than the real-life entity. Simulations prove to be very beneficial when experimenting with limited resources and data is being used from multiple sources.

   The real-time data aspect of DTs allows the use of predictive analysis in order to "predict" if the model will fail or how to achieve the best results.
4. Improving trust

   DTs can improve trust in the system because it gives a global view of security, reliability, and availability.
5. Enabling collaboration

The use of a DT will allow collaboration across different teams and sectors within an organization to work and collaborate seamlessly.

Types of DT patterns XMPro is currently using:

Status twin

- Used for basic condition monitoring
- Indicates operating parameters
- Created using visualization tools

Operational twin

- Used in decision support by operators and other decision-makers
- Used for actions where users can interact with the DT and change the operating parameters

Simulation twin

- Simulation capabilities to predict, forecast, and provide insights to future states
- Used for predicative maintenance or to improve recovery yield

How to use AI and machine learning with DT

- Real-time analytics
  - In the simultaneous data intake of DT, it can use both AI and machine learning to look for anomalies in the data, predict future situations, and optimize production.

- Decision support
  - DTs can be further used to display future predictions. This will help engineers to make more efficient decisions in the real time knowing the possible future consequences. This will allow faster decision-making in real time in response to critical situations.

- Prescriptive analytics and recommendations
    - DTs can be used for prescriptive analytics to create reconditions, the best next action to take based on the predictions.

8.5.1 Combining blockchain and digital twins

Using DTs with blockchain revolves around storing information from the DT and how it interacts in the blockchain. The second use is to allow the DTs to develop smart contracts with other digital systems.

Benefits of storing DT data in a distributed ledger are as follows:

- No central organizations that manage the information
- Lowered risk of tempering because of validation requirement
- Easy transfer of information while maintaining integrity

8.5.2 Smart contracts

Code that runs in a ledger when two or more bodies agree on a rigger event.

Area of opportunity for DT in industrial scenarios, e.g. they can interact with smart contracts in real time and execute commands in the code. The DT also has the capability of running analytics to the data in order to decide when to execute a smart contract in the business rule sense. Micropayments and virtual currencies are other uses emerging from the use of DTs and blockchains. Even though some of these may be explored further into the future, there is an immediate opportunity for DTs and smart contracts in the current sense.

Types of data to be used in the creation of DTs:

- Physics-based models (FEM, thermodynamic, geological)
- Analytical models (predictive maintenance)
- Time-series data and historians
- Transactional data (ERP, EAM)
- Master data (EAM, AF, BPM)
- Visual models (CAD, AR, VR, BPM, BIM, GIS, and GEO)

The selection criteria of the data are based upon the problem that the DT is being initiated to solve. It is possible to combine data from multiple types to create a suitable DT to create a holistic, real-time view of the entity.

8.6 Digital twins and heterogeneous data sources

A vital step of the DT creation is to create a link between the different data sources for the DT and understand the intertwined relationship between them. The relationships can be linear but additionally a graph-based structure can be used to observe the relation between the data and how it relates with the entity.

8.7 Digital twin versus digital thread

A DT is a DT representation of an entity. It stores historical and operating information and can perform analytical operations to provide recommendations to make real-time decisions. The DT provides a dashboard that indicates how the entity is preforming in real time and how it may perform in the future based on decisions taken.

However, a digital thread shows the birth to death records of the entity as it goes through the lifecycle. Whenever something of significance such as critical maintenance or anomalous operational behavior occurs in the lifecycle of the entity, it will create a memory and add it to existing records. In other terms, when an entity is retired, the owner will have a complete birth-to-death record of the entity in the form of a digital thread.

8.8 Creation of digital twin applications

XMPro Digital Twin Reference Architecture

- Mainly focused on DTs for owner-operators in asset intensive industries
- Focused on operational and continuous intelligence applications

Three main areas of the XMPro Architecture:

1. Data abstraction
   (a) Collection, wrangling, and storage data from different sources.
2. Analytics
   (a) Business rules, algorithms, and simulations to create meanings and trends from data.
3. User Interface
   (a) Where the user interacts with the DT through dashboards, workflows, etc.

Recommended steps for building a DT:
Step 1: Definite the function of the DT

Decide what you want your DT to do and what data you will need to make the decisions. Then you have to decide what actions will be taken based on those decisions. The best way to build a DT is to build it around a specific challenge.

Questions to ask while planning:

- Does it need to continuously monitor something and let you know when an event occurs?
- Should it provide decision support or decision automation, where it applies business rules or looks for anomalies?
- Can it apply machine learning or artificial intelligence? Phrased another way, if your DT was a real person, would you like them to use their intelligence and interpret the data?
- Should your DT make a recommendation on what to do next, or would you want it to take action on your behalf?

Step 2: Design the real-time data flow

In this step, the user can connect the different data sources that are to be studied and add the different analytical operations and workflows. Once that has been set up, the visual aspect can be created.

Step 3: Create the visualizations and actions

The user will have a real-time visualization of the data at this stage so that they can observe what and why things are happening, enabling them to make decisions and recommended actions.

Step 4: Iterate and improve

Once the DT has been deployed, there will be a lot of learning about the data that is being input and output, often even unexpected things. Moving further with the DT users can make changes to improve and adapt with their DT to get the maximum use out of it and its capabilities.

References

[1] Z. Guo, J. Zhang, R. Zhang and X. Zhang, "Aviation-to-Grid Flexibility through Electric Aircraft Charging," *IEEE Transactions on Industrial Informatics*, vol. 18, no. 11, pp. 8149–8159, 2022.

[2] A. Barzkar and M. Ghassemi, "Electric Power Systems in More and All Electric Aircraft: A Review," *IEEE Access*, vol. 8, pp. 169314–169332, 2020.

Index

www.ingramcontent.com/pod-product-compliance
Lightning Source LLC
LaVergne TN
LVHW011852200926
848484LV00002B/441
* 9 7 8 1 8 3 9 5 3 7 2 1 9 *